A Journey with
Fred Hoyle

Second Edition

T0222478

A Journey with
Fred Hoyle

Second Edition

Chandra Wickramasinghe

University of Buckingham, UK

edited by Kamala Wickramasinghe

World Scientific

NEW JERSEY · LONDON · SINGAPORE · BEIJING · SHANGHAI · HONG KONG · TAIPEI · CHENNAI

Published by

World Scientific Publishing Co. Pte. Ltd.

5 Toh Tuck Link, Singapore 596224

USA office: 27 Warren Street, Suite 401-402, Hackensack, NJ 07601

UK office: 57 Shelton Street, Covent Garden, London WC2H 9HE

British Library Cataloguing-in-Publication Data
A catalogue record for this book is available from the British Library.

A JOURNEY WITH FRED HOYLE
Second Edition

ISBN 978-981-4436-12-0 (pbk)

Typeset by Stallion Press
Email: enquiries@stallionpress.com

Printed by FuIsland Offset Printing (S) Pte Ltd Singapore

To Priya, whose steadfast support was a source of supreme strength

Foreword

This expanded second edition of a "Journey with Fred Hoyle" brings up to date a scientific story of enormous historical importance. Astrobiology is one of the most rapidly growing scientific endeavours of our times. Space exploration in the foreseeable future may soon uncover irrelutable evidence that the late Sir Fred Hoyle and Chandra Wickramasinghe, who heads the Centre for Astrobiology at the University of Buckingham, have been correct all along in their radical theory that terrestrial lifeforms are not unique but microparts of one stupendous cosmic whole.

Lord Tanlaw
Chancellor: University of Buckingham

Foreword to First Edition

This is an autobiographical account of a remarkable collaboration between two of the best known astronomers of our time — Chandra Wickramasinghe and the late Sir Fred Hoyle. Chandra's early life in Sri Lanka forms the backdrop for a remarkable scientific contribution of the 20th century. Investigations of the composition of cosmic dust begun in the 1960s led Wickramasinghe and Hoyle to the unlikely conclusion that the Universe is teeming with microbial life, and that such life can be transported from one cosmic location to another. Seen from this point of view, we are part of a connected chain of being that extends from the Earth to the far reaches of the Universe. Science fiction writers have for decades talked about our genes coming from stardust. Diligent and painstaking research over 30 years has turned a once heretical idea into mainstream science. It is a case of science fiction turning into science fact.

Ironically, Sir Fred developed this idea in his science fiction novel *The Black Cloud*. What a pity he did not live to see its possible realization.

Sir Arthur C. Clarke (1917–2008)
Fellow, King's College, London

Preface to Second Edition

It is now 50 years since our first paper proposing a carbon model of interstellar grains was published, and a full 35 years since our papers introducing interstellar biochemicals and the background to panspermia were published. Our graphite grain model of 1962, that was so violently opposed at the outset, is now part astronomical culture — every astronomer working on grains includes a graphite component, invariably forgetting its original provenance. While evidence for the widespread occurrence of interstellar organic molecules that can be called biochemicals is also now overwhelming, there still remains a resistance to regard them as degradation products of biology. Fred Hoyle and I argued that this was the only option consistent with all the information we possessed — from biology, geology and astronomy. Indeed the synthesis of these disciplines leading to the new subject of astrobiology was spearheaded by Fred Hoyle in the early 1980s. Our contention is that the *first* origin of life in the Universe involves the overcoming of well-nigh insurmountable odds, but its subsequent survival and spread across galaxies via processes of panspermia is comparatively trivial and indeed inevitable. While panspermia still lingers on the edge of orthodox science, the flow of strongly supportive evidence seems hard to stem.

Many universities around the world now include astrobiology in their academic portfolios. Several space missions related to astrobiology have been conducted, and others are on the way. The European Space Agency's *Rosetta Mission* is due to make a landing on a comet in 2014 and to conduct the most extensive set of experiments to probe its surface. NASA's Curiosity Rover is rolling over the plains of the Gale crater of Mars and over the next few years will look for signs of past or present life. A re-evaluation of data from the earlier 1976 Mars *Viking* probe is providing startling evidence to support Gil Levin's original attestation that microbial life was already discovered at the Viking landing site. An astrobiology mission to the three icy moons of Jupiter was approved by ESA in May 2012. By the year 2030 we can confidently expect to have confirmation of microbial life in the subsurface oceans that are thought to exist on these satellites. NASA's *Kepler Mission* has discovered several Earth-mass planets on which life could survive, including the planet Kepler 22b located 600 light years away in the constellation of Cygnus. The search for similar planets continues today at a hectic pace.

The idea that life is everywhere in the cosmos has gained ground to the point that Fred Hoyle's 1980 prediction can now be seen to be amply fulfilled:

> "... the cosmic quality of microbiology will seem as obvious
> to future generations as the Sun being the centre of our solar
> system seems obvious to the present generation ... "

Chandra Wickramasinghe
Cardiff, UK

Contents

Prologue

Harmony:
The stars shine
I gaze at them.

Among a myriad stars
I stand alone

And wonder
How much life
And love
There was tonight.

Written by the author in 1956

After Fred Hoyle had published his second volume of autobiography entitled: "Home is where the wind blows" he told me on one of his last visits to Cardiff:

"I have made only a passing reference to our long collaboration, because it seemed disjoint ·from the thesis I was developing there. Ours is an even bigger story that is certainly worth telling. Perhaps you would like to do that some day?"

The journey described in this book leads to an inescapable truth about the origins of life — humans, animals, plants and indeed all life on Earth share a cosmic ancestry. Our genetic heritage is derived from the wider universe; our genes neatly packaged within bacteria came from space.

This radical position was by no means rashly conceived. The long series of steps, over four decades, in reaching it were taken with utmost caution and always with a measure of trepidation. Every single unorthodox step was taken only after more conservative alternatives were carefully evaluated. A suite of ideas concerning the cosmic nature of life that was considered outrageously heretical twenty-five years ago, is now sliding imperceptibly into the domain of orthodox science.

Yet the word "panspermia", the concept that life — actual and potential — pervades our universe has its origins in classical Greece and dates back to the pre-Socratic philosopher Anaxoragas (c 500–428 BC) who posited that living seeds or "spermata" have been ever-present in the Universe. In India, in Vedic traditions that go back still further (c 1200 BC), the notion of life being an ever-present cosmic attribute is well-chronicled. These ideas had little impact in the development of Western philosophy, however. The philosophy of Aristotle (c 384–322 BC), based on empiricism, gave the Earth a position distinct from the rest of the Universe, which was thought of as an intangible abstract concept not amenable to reasonable study. Aristotelean ideas that dominated Western thought for centuries included the doctrine of the spontaneous generation of Earth-based life. Fireflies, for instance, were said to emerge from a mixture of warm earth and morning dew, a fanciful explanation which was nonetheless held as irrefutable fact.

The idea of spontaneous generation received its first serious challenge in the 1860's with the work of Louis Pasteur (1822–1895), which suggested that even the very simplest known organisms could not arise independently of a parent organism. This led to the view that each generation of every plant or animal present today must be preceded by an earlier generation of that plant or animal. The same causal link, including the effects of evolution, must persist all the way through the record of fossils and microfossils on the Earth, to the moment when the first microbial life appears. The fact that we now know this moment to have occurred during an epoch of intense cometary bombardment some 3.8–4 billion years ago, possibly means that the life-from-life connection could be extended to a time before the Earth itself existed. This logical outcome of Pasteur's life-from-life paradigm was recognised by several distinguished scientists in the latter part of the nineteenth century, notably John Tyndall, Lord Kelvin and Hermann von Helmholtz. Helmholtz' succinct summary of his position speaks eloquently for them all:

> *"It appears to me to be fully correct scientific procedure, if all our attempts fail to cause the production of organisms from non-living matter, to raise the question whether life has ever arisen, whether it is not just as old as matter itself, and whether seeds have not been carried from one planet to another and have developed everywhere where they have fallen on fertile soil..."*
>
> (H. von Helmholtz in *Handbuch de Theoretische Physik*,
> Vol. 1, Braunschweig, 1874)

Despite this early advocacy of panspermia, it was Svante Arrhenius (1859–1927) who first elaborated upon the idea in a quantitative way, and adduced evidence for survival of spores, whilst also proposing an explicit mechanism for interstellar transport (Svante Arrhenius, *Worlds in the Making*, Harpers, London 1908).

Arrhenius' ideas quickly fell into disfavour partly because it was felt that microorganisms could not be expected to withstand space conditions, and partly because the theory was regarded as intrinsically untestable. The former objection has turned out to be largely false — bacteria are incredibly space-hardy — but the latter point concerning lack of testability remains an issue in Arrhenius' rendering of panspermia. The claim that four billion years ago life on Earth evolved from a primordial seed or spore from space is not testable. But if it could be demonstrated that living seeds still reach the Earth from space, the essence of his theory would be vindicated. Arrhenius asserted that the process he described would be exceedingly difficult, if not impossible, to detect. Science does not look kindly on untestable hypotheses, so Arrhenius' version of panspermia was destined to fall on fallow ground.

Our own approach to the theory of cosmic life, seventy years later, led us at every stage to propositions that were imminently testable or verifiable. Very many of our predictions have indeed been verified, and a multitude of other tests are now in progress.

Chapter 1

Origins: Prelude to the Journey

"There was nothing in the idea of evolution; rock pigeons were what rock pigeons always had been"

Wilberforce, 1860

During the debate on evolution which took place at a meeting for the British Association for the Advancement of Science in Oxford, June 1860, Bishop Wilberforce turns to Thomas Huxley: "Sir, is it on your grandfather's or your grandmother's side that you claim descent from a monkey?" Famously Huxley replied that he would sooner have an ape for an ancestor than accept the dogma of the church. This much mythologised altercation took place a year after the publication of Darwin's *Origin of Species*. Up until this time, the Judeo-Christian account of creation had dominated Western thought and Darwin's impacting new theory posed a significant threat to the Anglican Church, establishing the autonomy of Science. Darwin's theory of evolution grew in strength throughout the early 20th century with expressions of it being evident in literature, philosophy and social policy. Some sectors of the Church were to assimilate the secular aspects of his theory into their doctrines, but others remained staunchly opposed and an undercurrent of dissension remains even in the present day.

All thinking people were required to have an opinion on the subject of creation and even as late as the 1920's and 1930's students seeking admission to Cambridge University had to satisfy the examiners in a paper called "Paley's Evidences". My first sight of such

1

papers came from my father's documents — he was a Senior Scholar in Mathematics at Trinity College Cambridge from 1930 to 1933. William Paley (1743–1805) was a theologian and philosopher whose influence prevailed over the intellectual world for over two centuries. In his treatise *Natural Theology*, Paley introduced the renowned image of the "watch (that) must have a maker" to expound a teleological argument for the existence of God. The tenor of his work was almost scientific and quite distinct from the doctrines of a miraculous creation. He looked at the intricate workings of the natural world and conjectured that such perfection must imply the work of a Creator: "The marks of design are too strong to be got over. Design must have a designer. The designer must have been a person. That person is GOD".

Darwin was familiar with the writings of Paley, "the logic of...Natural Theology gave me as much delight as did Euclid", and his own works echo Paley's style and methodology. Both men sought to explain evolution in terms of mechanistic, natural processes, albeit to different ends. Where Paley would always revert to metaphysics to explain the first cause, Darwin remained elusive. Indeed in this respect the title of his *Origin of Species* is somewhat misleading, as he traces the evolutionary processes that led to the fact of Man, but seems to slide away from looking into the causes of the first stirrings of life. However, in a landmark letter to Hooker in 1871, he writes:

> *"It is often said that all the conditions for the first production of organisms are now prevalent which could ever have been present. But if (and oh! What a big if!) we could conceive in some warm little pond, with all sorts of ammonia and phosphoric salts, light, heat, electricity, etc. present, that a protein compound was chemically formed ready to undergo still more complex changes. At the present such matter would be instantly devoured or absorbed, which would not have been the case before living creatures were formed."*

It is interesting that Darwin, who never relished the prospect of being viewed as anti-establishment, chose the intimate medium of a letter to put forward his views on the origins of life. And the idea expressed here marks the beginnings of the primordial soup theory of life's origins that came to govern scientific thinking throughout much of the 20th century.

By 1939, the year of my birth, the primordial soup theory had gained so much credence that it was the only scientific way of looking at the problem of the origin of life. Indeed science proudly considered it to be a problem solved. The scientists Oparin and Haldane had set the ground rules for discussions of this theory that life on Earth originated on Earth. Recognising that the production of organic molecules in the present-day atmosphere of the Earth was unlikely, Oparin and Haldane argued for a hydrogen-rich primordial atmosphere in which the organic molecules required for the origins of life would be formed. The process first required the break-up of inorganic gas molecules such as water, methane and ammonia into reactive fragments or radicals through the action of solar ultraviolet light and electric discharges. Next, the radicals recombine through a cascade of chemical reactions, and in this process a trickle of organic molecules is formed. Such molecules, which are the chemical building blocks of life, then rain down into the primitive oceans forming a dilute primordial soup. It is from such a soup, through a multitude of chemical reactions over millions of years, that life is supposed to have begun.

This model of life's origins had rapidly acquired popularity and kudos in the post-Darwinian era. Haldane, a distinguished geneticist, physiologist and philosopher was reputed also to possess great charisma which he employed successfully to popularise his theory which eventually came to be written into biology textbooks as irrefutable fact. The reasons for its success were multifarious — it reassured man to consider himself to be autonomous with terrestrial origins, and in an age where religious beliefs were dwindling, Science was stepping in to offer man a "logical" explanation of his origins. The only problem with the primordial soup theory was that it had no foundation in fact. In the 1920's and 1930's, there was no empirical basis whatsoever for its justification, and a judgement had to be made solely on aesthetic or philosophical grounds. For the Soviet scientist Oparin, an overall consistency of this theory with

Marxist materialistic ideologies appears to have been an important consideration. Haldane too had espoused Marxism in the 1930's and was for several years editor of the *Daily Worker*. In Haldane's chequered career he later became disillusioned with communism, as indeed with British imperialist policy, and in 1957 emigrated to India where he spent the rest of his days.

Retrospectively, I suspect that Haldane's communist leanings may have contributed to the alacrity with which Fred Hoyle embarked on our challenge of the Oparin–Haldane model in the 1970's. Fred Hoyle, though he came from working class parentage — his father was a wool merchant — was always a staunch supporter of British Conservative politics. His distrust of communism and Marxist philosophy may have added to his suspicion of theories — even scientific theories — that had sprung from such a system of thought. But his political leanings quickly became incidental in our opposition to the received theory of the origins of life — our bold refutation emerged as a natural outcome of our fervent interests in astronomy.

Whilst scientists in the modern age like to believe that their activities are always free of prejudice, such a position cannot be further from the truth. At the deepest level, science, particularly when it comes to fundamental questions such as the origins of life, is inextricably linked to cultural traditions. That includes political as well as religious prejudice. Although subconsciously ignored or sublimated they remain as invisible constraints.

Some aspects of my personal life are relevant to the thesis of this book only insofar as they connect to the remarkable story of my journey with a man who was amongst the most original and imaginative of scientists in the 20th century. My collaboration with Fred Hoyle from 1960 onwards led me, over four decades, to question one of the most cherished paradigms of science. We did not meet until the 21st year of my life. But the period of my life up to this time that was spent in my native land of Sri Lanka was a preparation for the unique adventure that was to follow.

My early years spent in Sri Lanka followed a more or less predictable course *vis-à-vis* my circumstances. Sri Lanka (Ceylon as it was then called) was an outpost of the British Empire, in many ways overshadowed in importance by the neighbouring subcontinent of India. It is a fertile island of supreme versatility accommodating mountains, rain forests and beaches within its relatively small confines. The interest it held for its successive colonisers was, however, mostly commercial, in its precious stones, spices, coffee and later tea. The Dutch, the Portugese and the British prized this colony mostly for these commodities as well as for the strategic location of its natural harbours on sea routes to the Far East. Sri Lanka has a history stretching back over two millennia with sprawling sites of ancient ruins testifying to epochs of past splendour. However, four centuries of colonial domination had to a great extent left Ceylon demoralised in a state of national lethargy, and even its struggle for independence from Britain, which was achieved in 1948, was a pale shadow of the emotions expressed in the independence struggles of India.

In my earliest recollections Ceylon was an impoverished feudal society with a sharply visible division between rich and poor. The rich privileged class had access to good schools, whilst the poorer underclass was only minimally literate with limited access to education. The country was also sharply divided between the amenities available in a few major cities (e.g. Colombo, Kandy and Jaffna) and a multitude of villages of which the country was comprised. My own home was in the capital Colombo, and my school Royal College Colombo, established in 1838, was modelled on the traditions of the English Public School system. My teachers, specially in mathematics and physics, conveyed to me their own passion for these subjects, and were a source of inspiration in my formative years. I was lucky too in that my father was a talented mathematician who obtained the highest honours in this subject both in Ceylon and in the Mathematical Tripos in Cambridge, where he became a B star wrangler in 1933. Not only did I have this added source of stimulus at home but I was also surrounded by my father's collection of mathematical and astronomical books which included classics such as Eddington's *Mathematical Theory of Relativity* and Brown's *Lunar Theory*, not to mention an extensive popular list.

A benefit of living in Ceylon in the 1950's was that the environment was still pristine and unpolluted. There were no bright street lights in the suburb where we lived and hardly any pollution from cars and buses, so that the pageant of the night sky was magnificently brilliant. We lived close to a beach by which a railroad ran connecting Colombo with smaller cities in the south. I would often walk along this beach in the evening, sometimes along the railroad sleepers, and watch the sun set over the Indian Ocean. I vividly recall my childhood experience of spectacular sunsets such as I have never since seen. Within minutes the sunset disappears into a wide black canvas overhead studded with millions of stars. Looking up at the myriads of stars that populate the Milky Way, contemplations about man's place in the universe were inevitable.

It is rare to see such a spectacle nowadays in our modern cities with their deplorable output of light pollution. Our inability to enjoy our natural heritage of the night sky leaves us far poorer, and also less able to make a connection between ourselves and the wider cosmos — a connection that was deeply felt by our ancestors.

Sri Lanka is steeped in Buddhist traditions and its influence is inescapable. The island is strewn with ancient temples and 2000 years of Buddhism literally permeates the land. Buddhist descriptions of cosmology that date way back to the early Christian era are distinctly post-Copernican. In a Buddhist text *Visuddimagga* (written in Sri Lanka in the 1st century AD) it is stated that:

> "... *as far the these suns and moons revolve shining and shedding their light in space, so far extends the thousand-fold universe. In it are thousands of suns, thousands of moons ... thousands of Jambudipas, thousands of Aparagoyanas ...*"

the latter being translated as meaning extraterrestrial abodes of life. The billions of galaxies of modern astronomy could be identified in statements found in other contemporary Buddhist texts which referred to the entire Universe as "this world of a million, million world systems". Such passages made a significant impact on me in my young years, and I noticed a striking similarity between these ideas and the ones expressed by James Jeans in his *Mysterious Universe*.

My resolve to study astronomy was strengthened by an astronomical event that was fortuitously connected with my homeland.

A total eclipse of the Sun, visible from Sri Lanka, was to take place on June 5, 1955. I was 16 at the time and my serious interest in science was just beginning to develop. Sri Lanka, which had hitherto been a scientific backwater, was suddenly transformed into a hive of professional scientific activity. This particular eclipse was to have the longest period of totality since AD 699 and several important scientific experiments were being planned. Scientists from Britain, USA, France, Germany and Japan all converged here and the local newspapers were full of news about these momentous scientific events. One experiment that was planned was a test of Einstein's Theory of General Relativity which predicts a bending of light by a small predictable amount (1.75 arc sec) as the light of a star passes close to a massive object like the sun. The project was designed to validate an experiment of a similar kind carried out by a team led by Eddington during the solar eclipse of 1919.

As a keen amateur photographer, I had set up my own experiment with a simple camera fixed at the end of a home-built telescope to capture the event. We were of course warned about the dangers of looking directly at the Sun, so at the appointed hour, we had our darkened glasses and basins of water in place to watch the progress of the eclipse. I watched with bated breath as the Moon slid ominously over the Sun's disc, casting an instant gloom over the landscape as in an impending thunderstorm. Then total darkness descended suddenly, lasting for an interminable seven minutes. There was a noticeable chill in the air and a denatured atmosphere. Lotuses began to fold their petals inwards, animals cowered and crows cawed wildly. The fabled spectacle of the solar corona with its outstretching flaming tongues was visible intermittently through transient clearings in a thin veil of drifting cloud. Then it was all over, the noon day sun mystically reappeared. I felt more than ever before the indomitable power of the cosmos.

As often happens with observations in astronomy, most of the observing teams in 1955 were disappointed because clouds intervened, but a few stations were able to make successful observations that led to new science. To a teenage admirer of science these events provided a thrilling experience. Science was happening at my own doorstep, and the events of 1955 played no small part in my determination to pursue astronomy as a career.

Nowadays most students enter astronomy through undergraduate courses in physics, or physics and astronomy. When I asked around for advice on how one became an astronomer the answer I was given by informed persons, not least my father, was through mathematics. So I entered the University of Ceylon in 1957 as an entrance scholar in Mathematics and had set my sights on my father's old University, Cambridge, should things work out for me as I had hoped.

The transition from school to University was an easy one as I attended University from home. The University of Ceylon, Colombo was only 20 minutes away by bicycle or ten minutes by car. I took my three years of University studies in my stride, following most of the courses in applied mathematics, because I felt this was the most important tool for exploring the Universe. I was lucky to have some excellent teachers who inspired me. A person who influenced me greatly in those early years was the professor of mathematics C.J. Eliezer who was himself a distinguished Cambridge product, a former Fellow of Christ's College and a pupil of the illustrious physicist Paul Dirac, whose major success was to reconcile special relativity and quantum mechanics. Through Eliezer's lectures I obtained exciting insights into the theory of electromagnetism, a subject in which I was later to specialise. I did not realise at the time that Eliezer and Fred Hoyle were Cambridge contemporaries and that they both had associations with Dirac. Because of the connection between these three people — Dirac, Eliezer and Hoyle — it turned out by a really curious coincidence that Hoyle was to be the external examiner in mathematics for the University of Ceylon in the same year that I sat my final degree examination. It amused me later on to think that Fred would have read my examination scripts long before he ever set eyes on me.

In the summer of 1960, I graduated with First Class Honours in Mathematics and was awarded a Commonwealth Scholarship by the British Government to pursue postgraduate studies at Cambridge University. I applied for a place to do a PhD in Theoretical Astronomy at Trinity College, my father's old College, and I was delighted to be accepted, and more so to be told that I would be supervised by Professor Fred Hoyle of St John's College who was Plumian Professor of Astronomy and Experimental Philosophy at the University of Cambridge. Whilst at the University of Ceylon I had already read two classic books by Hoyle: *Nature of the Universe* and *The Frontiers of Astronomy*, both of which had made an indelible impression on me. So when I received a handwritten letter from Fred Hoyle at my home in Colombo, recommending a list of books to read prior to coming to Cambridge in October 1960, I was naturally overjoyed.

Chapter 2

Cambridge and a First Meeting

In September 1960, I found myself preparing to leave home for the first time. In those days the normal way to travel from Sri Lanka to England was by ship. Although air travel was rapidly coming into fashion it remained a considerably more expensive option reserved mostly for business travellers and the rich. On a warm September evening I sailed away from the port of Colombo aboard the P&O Liner *SS Orcades*, wistfully watching a palm-fringed coastline recede slowly into the distance. A two-week voyage took us through the Suez Canal, via Naples, Gibralter and Marseilles to Southampton. Nowadays such a voyage would be regarded as a luxury cruise. But my enjoyment of this new experience was hindered by sickness due to rough seas. I spent a lot of the journey in my cabin feeling sorry for myself whilst reading and re-reading Fred's *Nature of the Universe*, a book I still regard as one of the greatest classics of popular science.

We docked in Southampton on a cold grey autumn morning and I was met on board by a lady who was a representative of the Commonwealth Scholarship Commission who had sponsored my visit to the UK. This was the first year of awards in the Commonwealth Scholarship Scheme and the authorities were very much at the stage of finding their feet as to what should be done. I recall being herded along with students from other Commonwealth countries to an initiation week in the capital, London. Many of the aid programs sponsored by affluent Western countries to assist the development of the Third World are conceived with trade-benefits in view, and many set up to conscience cleanse for past colonial exploitation. I was more than a trifle surprised at the Commission's ignorance of the needs of their clientele from the Commonwealth. They presumed that all the beneficiaries of their scholarships came from African villages or from rural areas of India or Sri Lanka. Most of the advice proffered as to how to make the transition from ex-colony to the hub of the empire, was hardly relevant to myself or, I suspect, the vast majority of the Commonwealth Scholars who came from more or less Westernised backgrounds in their home countries.

The literature of England enthralled me throughout my childhood and adolescence. And the history of the British Isles, its culture, landscape and even details of place names, now coming into context, contained a dreamlike familiarity for me. What I found difficult to enjoy, however, was the climate — inclement weather or persistent grey skies, that ice-breaker of conversation among the English. The long dark evenings of winter still, after 40 years, make me yearn for a more equable tropical climate!

My first impression of London, despite the magnificence and grandeur of its history and its buildings, was one of alienation. My own personal sense of detachment from the people and the environment did, however, seem to mirror a more general malaise. The density of people, the frenetic pace of life, the catacombs of the tube and the pollution all added to my sense of estrangement. Such highly subjective early impressions altered only much later as I came to discover the theatre, music and art galleries that make London unique.

Cambridge, on the other hand, had an instant appeal. I arrived at Cambridge station on a brilliant autumn morning, and following the instructions I had received in London, took a taxi to the Porter's Lodge of Trinity College for my first meeting with Dr. Robson, Tutor for Graduate Studies at 11 am. I recall vividly the sense of awe that swept over me as I walked across the Great Court of Trinity surrounded by the ghosts of legendary figures in both science and literature. This was the College of Isaac Newton, Bertrand Russell and William Wordsworth, and a brochure I had in my hand directed my attention to Newton's rooms above the porter's lodge, where his classic experiments on prisms and light were done.

Dr. Robson's welcome was eminently professional. He made me feel relaxed with a glass of sherry and offered a personal touch to our meeting by talking about my father's distinguished record as a student in 1930–1933. This, I later discovered when I myself became a Fellow and Tutor, was a trick that had to be learnt! After my first lunch in the College Hall and a few other meetings with Junior Bursars and the like, I arrived at my lodgings (a room) in a Trinity College house for graduate students in Burrells Field. Among a pile of inconsequential letters that awaited me was a handwritten letter by R. A. Lyttleton (Ray Lyttleton) of St. John's asking me to see him in his rooms to discuss matters relating to my supervision. That surprised me somewhat because I had already received a letter from Fred Hoyle in Sri Lanka saying that he would be supervising me.

When I met Lyttleton at the appointed hour I learnt that Fred Hoyle was in the USA for most of the Michealmas term and that he, Lyttleton, would supervise me to start with. My encounter with Lyttleton was a little short of formidable as I now remember it. After inquiring whether I had any specific problems in mind that I wished to work on, he informed me that astronomy was an extremely difficult subject to research. Most of the easy problems had already been solved, and what remained unsolved were insurmountably difficult! Only much later did I come to realise that these remarks were probably true to an extent if one confined one's attention to areas of classical astronomy that excluded applications of physics, astrophysics as it is now called. At this first meeting he asked me to take a look at a rather abstract problem in the theory of stellar structure concerning "solutions of Emden's equations for polytropes". He also very generously gave me a copy of his book on comets (which I still possess), in which he described the dust bag theory of comets, following from work he had done in the 1940's. The book was in fact an elaboration of a paper he co-authored with Fred Hoyle on the accretion of interstellar dust by the sun. Lyttleton, it should be noted, was the person who turned-on Fred's interest in Astronomy at a time when he was beginning to find Nuclear Physics a fallow field of research in the 1940's.

Apart from Fred Hoyle and Hermann Bondi, Lyttleton had hardly any other collaborators or graduate students as far as I know. Shin Yabushita, who arrived in Cambridge a year after I did, became his student, but it appears that Yabushita went his own way receiving little or no guidance from Ray Lyttleton. It was clear at the outset that I was not going to get much guidance from Ray, and I do not think I saw him again during my first few years in Cambridge. I had enough reading to get on with so I decided to bide my time until Fred Hoyle returned from the States. I also attended a few advanced lecture courses that were being offered to students of the Mathematical Tripos. These included Paul Dirac's course on Quantum Mechanics and Mestel's course on Cosmic Electrodynamis. All this was in preparation for when Fred Hoyle would return from the USA.

The next significant event I remember from my early Cambridge days was an invitation to tea by Jayant Narlikar. This was my first meeting with another of Fred's students who started research in the academic year 1960/61. Jayant, who later became a good friend of mine, was a star pupil in the Mathematical Tripos that summer (he was a star wrangler and a Tyson Medallist in the Tripos). When I met him in October, Jayant had already started serious work on a research problem in cosmology. Fred Hoyle had clearly spotted his exceptional mathematical talent and set him on a course to assist him in a long and bitter cosmological battle that lay in the years ahead.

Jayant and I shared a hut that was the temporary home of Theoretical Astronomy which was an outpost of the Department of Applied Mathematics and Theoretical Physics. My conversations with him made it clear that the 1960's were turning out to be a watershed era for cosmology. A right royal battle between the Steady State Theory of Hoyle, Bondi and Gold and the rival Big Bang Theory was set to begin. Cambridge Radio astronomers led by Martin Ryle had been claiming in the late 1950's that the Steady State Theory could be disproved by their study of the distribution of galaxies in the universe emitting radio waves (radio source counts). Hoyle claimed that these arguments were far from secure, and indeed were most likely to have been contrived. It turned out that Ryle's early analysis of the matter was based on limited surveys and consequently poor statistics.

Throughout successive surveys of radio sources made at the Mullard Radio Observatory in Cambridge, Ryle and his team pressed home the point that there was enough evidence to disprove Steady State Cosmology. In a simplistic interpretation, the universe appeared, on the basis of source counts, to be more compact at the earliest epochs, favouring the concept of a Big Bang origin. Jayant Narlikar came on the scene amid gathering storm clouds, and the atmosphere in Cambridge was highly charged when I first arrived.

Fred Hoyle had set Jayant the task of re-examining Ryle's radio source count claims before he had gone to the States earlier in the Summer. Without embarking on an extended diversion on cosmological history I should say that this particular storm was eventually weathered. Radio source counts did not say anything very definite. But other conflicts were to follow. Together with Fred, Jayant had embarked on a valiant course of defending steady state cosmologies, an endeavour that was to occupy most of his professional career.

The insatiable appetite for denigrating Steady State Cosmology, often with the flimsiest of evidence, that I witnessed in the 1960's still puzzles me. I cannot help thinking that the reasons have a deeply cultural basis. Without going into the technical details on either side of the argument, my own cultural predilection was for a steady state universe of some kind. Such a cosmology is consistent with the philosophical world view that pervades the Indian subcontinent, and in particular it is in harmony with Buddhist traditions that are prevalent in Sri Lanka. Another aspect of the Cambridge debacle that astonished me was the component of jealousy that entered a scientific controversy. In my naivety I had believed that science was pursued in a cold detached manner independent of personalities or social constraints. This was far from the truth. It was clear from the situation I witnessed that the two contesters in this argument were worlds apart in their personalities and ideologies. Fred was a forthright and candid Yorkshireman; Ryle a stiff upper lip product of the public school system. Never would their differences be reconciled.

When my turn finally came to meet Fred Hoyle, problems of enormous gravity would have been on his mind. As I walked from my lodgings on a frosty afternoon in December, across the backs of the river Cam, over a desolate playing field of St. Johns, I wondered what my long-awaited meeting would be like and what it might eventually lead to. Would I encounter a man weighed down by the strain of a huge controversy, preoccupied and terse; or a sparkling communicator, the author of the some of the most stimulating books I had read? But my anxieties were quickly dispelled when I arrived at the door of 1 Clarkson Close.

Barbara Hoyle greeted me with an unprecedented warmth of affection and generosity. Fred, I was told, was just finishing an interview with a journalist and I was ushered in to their dining room to take tea with Barbara and her mother, Mrs Clarke. The ladies were so welcoming that in minutes I felt completely at ease, and within half an hour, that I had known them for a lifetime.

When the journalist had left I was taken into a spacious, thickly carpeted open plan living room that served as both library and study. Large patio windows extended over the entire length of the room on the far side, looking out on a slightly undulating lawn. There was Fred seated on his easy chair by the window, with a writing pad on his knee, fountain pen in hand scribbling a calculation with intense concentration. It was hard to decide whether he was pleased or not to be disturbed, but as Barbara introduced me as his new Ceylonese student, he switched into a more relaxed mode, stood up, shook my hand and uttered his standard greeting that all his students came to gleefully imitate, "Oh Hellow!"

I cannot remember exactly how our conversation went, but I do recall a remark that embarrassed me such as "I hear you write poetry?" He must have seen such a declaration on the application forms that were sent to the University. On confessing that I had published a slim volume of poems and had some poems included in an *Anthology of Commonwealth Poetry* that was published by Heinemanns that year, he was noticeably impressed. We had established a connection at a deep level — an abiding passion for creative writing and a love of the English language.

We talked about the Lake poets, about Malowe and Ezra Pound. We talked about politics, about cricket, about Sri Lanka and my old college professor there, who had been a Cambridge contemporary of

Fred's. We even talked about the weather. In fact we talked about everything except science. I took all this to mean that he had not yet given much thought to what I might do as a research project, probably because of his preoccupation with other more incumbent matters at the time. He did however direct me to his own monograph on Solar Physics and a book by Cowling on Magnetohydrodynamics, but without any explicit application to think about. He also suggested that I attended a Part II course on Electromagnetic theory and a Part III course on Stellar Structure both of which were to be given by Fred in later terms during the academic year. I left Clarkson Close that evening feeling greatly relieved and excited. And I had finally met one of my childhood heroes in science.

During the next two months I must have seen Fred on half a dozen occasions, always at home. The explicit problem to which I was first directed by Fred, was not in any of his main areas of concern at the time. It was the problem of the origin of the Sun's polar magnetic field and of its alternations of polarity (north pole becomes south and vice versa) through a 11 year solar cycle. Fred's idea was that the polar magnetic field was built up by small-scale magnetic loops carried outwards by streams of charged particles. My suspicion was that Fred had already worked through this model in his head, even to the minutest detail and required me only to confirm his hunches with a few straightforward calculations in electromagnetic theory. This I did with relative ease. It seemed strangely appropriate that my first research project should concern the sun's polar field given that it was a solar eclipse that had provoked my desire to become an astronomer. Within only a few months of our first meeting I found myself becoming co-author of my first scientific paper with Fred which was published in the Monthly Notices of the Royal Astronomical Society (F. Hoyle and N.C. Wickramasinghe, *Mon. not. R. Astron. Soc.* **123**, 51, 1962).

Chapter 3

A Hike in the Lake District

Seeing our first paper in print gave me a thrill, even though my own part in the project had turned out to be relatively minor. But although the problem of the polar field was interesting enough in itself, I could not muster enough enthusiasm to think deeply about it. I began to wonder how this problem with its rather limited scope could pan out into a substantial programme of research occupying the full three years of my Commonwealth scholarship.

I spent most of the Lent and Summer terms of 1961 profitably engaged in reading widely around my subject and attending a variety of lecture courses. I enjoyed the new freedom of being able to learn without the overhanging threat of examinations. This was also a time of readjustment to a new country and a different way of life for me.

And the way of life in the Western world at large also seemed to be undergoing rapid change. There was the expansion of global air travel and the advent of the concept of multinational industries. The stage was set for the modern world of instant communication, although the internet itself was still a couple of decades in the future. More importantly in relation to the subject of this book, we were on the threshold of the Space Age that was destined to transform astronomy for ever.

A less tangible transformation of social attitudes had also begun. The election of John F. Kennedy as President of the United States was seen as a new beginning. A quiet confidence and optimism began to pervade Western society. The lean post war years of the 1950's had given way to an unprecedented economic boom that had an effect on all our lives and a feel good factor was evident in people's psychology. Problems that were looming large on a distant horizon — the Cold War, the Arms Race, the growing dissension in South Africa, conflicts in the Middle East and the slow beginnings of global terrorism — made no impression on the day-to-day lives of people in the West. Along with slogans for combating communism came clamours for freedom, liberalism and permissiveness. *Lady Chatterley's Lover* found its acrimonious entry into the literary canon, and the cause of feminism was taken a step further with the election of the World's first woman Prime Minister in Sri Lanka. For me the new found sexual freedom that was given abundant expression on the streets of Cambridge came as a shock to my sequestered upbringing in an ex-colony with distinctly Victorian values.

The 1960's could also be seen as the halcyon days for Science. Most areas of research advanced steadily and with the same confidence that characterised society as a whole. The dawn of the new decade was marked by several notable technical triumphs. The laser, a light source of unprecedented intensity, made its debut in 1960. The quark theory proposed that particles like protons and neutrons, hitherto thought to be the fundamental units of matter, were in turn made up of even more basic units called quarks. Man walked on the Moon. The exploration of space was well under way.

All these developments seemed to flow naturally and with ease, but it should not be forgotten that the groundwork — perhaps the hardest groundwork — in many areas was already laid in the 1950's. It was in the mid-1950's that radioastronomy began, and amongst its rich harvest are discoveries that would turn out to be germane to the questions that will be addressed later in this book. For instance, the study of how hydrogen gas is distributed in the galaxy, and the discovery of molecules connected with life, all stemmed from developments in radioastronomy. The mid-1950's also saw the successful completion of the work by Geoff Burbidge, Margaret Burbidge, Willy Fowler and Fred Hoyle that led to an understanding of how the chemical elements, including the elements of life, are formed from hydrogen in the deep interiors of stars. At about the same time, there was the monumental discovery by James Watson and Francis Crick of the famous double-helix structure of our genetic material — DNA. Shortly afterwards Fredrick Sanger analysed the nature of a protein (insulin to be specific), showing its detailed sequence of constituent amino acids; and Harold Urey and Stanley Miller completed their classic experiments in which they showed how the most basic chemical building blocks of life might be synthesised from inorganic matter.

It might seem ironical that the burgeoning of liberal traditions in society had little effect in promoting a spirit of free and unfettered inquiry on the scientific scene. Although science in general, and astronomy in particular, continued to flourish, there was a decline in the development of new ideas. Newly invented experimental techniques generated a vast body of factual data. But there was precious little being done in the way of a critical re-appraisal of old hypotheses. The belief grew that the really important problems were either close to being solved, or that they were really so hard that we should not even begin to worry about them. In practical terms it was a case of turning out more of the same type of factual data, attempting to consolidate existing theories, seeking to tie up the last loose ends. An openness of mind required for a vibrant scientific culture was noticeably lacking.

Such then was the general backdrop against which my specific experience of research continued through the early part of 1961. As mentioned earlier Fred Hoyle was engrossed at the time with trying to resolve the radio source count conflict with Martin Ryle — a battle that was becoming ever more bitter. Fred, however, never slackened his determination to defend steady state cosmology, in the conviction that the alternatives were far less credible, and the data adduced in their favour either contrived or indeterminate. His collaborations with Jayant Narlikar on many aspects of Steady State Cosmology were gathering momentum at this time.

Although Fred saw cosmology as his prime battle field in the 1960's, his interests were by no means restricted to cosmology. His knowledge and experience across the entire field of astronomy and astrophysics was impressively encyclopaedic. By 1961, there were few areas of this subject that had not been embellished by his imagination and genius in some way. In the year I joined him in 1960, Fred took on a record number of research students — four in all. Besides Jayant Narlikar there were John Faulkner and Ken Griffiths who were set working on problems of stellar evolution, and Sverre Aarseth who was working on the gravitational N-body problem.

By working on a wide range of problems through his several students Fred's attention may have been diverted from the bitterness of the cosmological conflict. Another welcome diversion he had was an alternative career as a science-fiction writer. In 1959, a year before my arrival in Cambridge, he had already published his novel *The Black Cloud* which in a sense was a forerunner to his much later collaborations with me on life in the cosmos. I acquired a copy of *The Black Cloud* in the spring of 1961 and remember reading it avidly, particularly because I was able to connect the political nuances and intrigues described in the book with what I had seen enveloping Fred's life in recent months. The hard scientific content of the novel also impressed me enormously. His arguments for complex molecules acting collectively as intelligent entities intrigued me even before I began to work on organic molecules in space. Fred's science fiction novels were not merely plausible, they were entirely in the realm of what was possible according to what was definitely known (or *almost* definitely known) about the real world.

Fred often gave an impression of cold detachment and indifference when he came to dealing with people. But with his collaborators and students he always displayed a profound insight as to their temperaments, special aptitudes and capabilities. In the spring and summer of 1961, I was still groping to find my next problem to tackle, and thoughts about the nature of interstellar matter began to enter my head. I saw Fred a couple of times and recall expressing more than a passing interest in the scientific content of *The Black Cloud*. He informed me then that he had found it difficult to publish his calculations that showed the hydrogen molecule and other molecules to be abundant in interstellar space, and that his novel was a way of expressing these ideas at a time when science would not countenance them.

I discovered that he was at this time also engaged in another science fiction project with TV producer John Eliot. Together they were writing the script for *A for Andromeda* which became a highly successful BBC TV serial when transmitted in 1961 and 1962. Later on it was to become a cult classic of sci-fi fanatics. In the script, a newly built radio-telescope picks up intelligent signals from the constellation of Andromeda which are fatefully interpreted as instructions for building a gigantic supercomputer. Once built, the computer begins to relay the information it receives from Andromeda and the security of humankind falls under dangerous threat. By presuming the presence of intelligent life in a far flung corner of the Universe, albeit through the guise of science fiction, Fred was moving unmistakably towards some form of astrobiology as early as 1961.

When my interest in *The Black Cloud* matters became evident Fred directed me to a review article on "Interstellar Matter" by Jesse Greenstein who was at the time working at the Mt. Wilson and Palomar Observatories. Greenstein had discussed the composition of interstellar dust, considering the arguments for two separate classes of dust grain — dielectric icy particles and metallic iron particles. At that time, however, there was no clear idea as to how either of these grain types could populate interstellar clouds. My reading introduced me to many questions that seemed to be in urgent need of answering.

In my initial discussions with Fred at Clarkson Close it was clear that he was not happy with any of the theories that were around at the time. It was only later that I realised the precise reason for his disquiet. He was at work at the time with Willy Fowler on a fragmentation model for star-formation and needed to have certain opacity and volatitlity properties for the dust. The properties they required did not conform with any of the existing grain models.

My 40-year long journey with Fred Hoyle was now about to properly commence. I arrived at the door of 1 Clarkson Close on a balmy evening in the late summer of 1961 and, as usual, I am welcomed by Barbara Hoyle. Before taking me to see Fred I was, as usual, brought up-to-date the family's doings. Barbara seemed more socially disarming than usual when she announced that she had decided that I should accompany Fred on a walking holiday in the Lake District. Here I could experience firsthand the romance of the Wordsworth country I had read so much about, and, as she put it "you men could discuss your work".

Walking in the Lake District was Fred's way of escaping from the tribulations of academic politics that had lately plagued him. In the mountains of the Lake District he found repose and time to think. Here he would experience for a while what he saw to be the real challenges of life, those that a mountaineer had to face, struggles with the elements and the terrain. In this way the irrelevant squabbles in the cloisters of university would fade momentarily into insignificance.

Needless to say I was overjoyed at the prospect of joining Fred on one of his walking escapades. Poised beside him in his two-seater Sprite, we set off on our five-hour long journey. As we passed through mountainous landscapes, Fred would describe how this magnificent scenery had been gracefully sculptured by ice floes as ice ages came and went over long periods of geological time. He was thinking aloud as he often did when he felt relaxed in appropriate company, and listening to his passing thoughts was always an invigorating experience.

We arrived at the Old Dungeon Ghyll Hotel, then a modest guest house in Little Langdale (near Ambleside). It has since been transformed into an elaborate hotel, but in those days it was a serious hiker's retreat, run by the mountaineer Sid Cross and his wife. It was early evening and there was just enough time to unpack before we sat down to a sumptuous dinner cooked by Mrs Cross. The evening meal at the Old Dungeon Ghyll was the focal point of the day and was more than ample compensation for its somewhat spartan accommodation.

After browsing through the day's newspapers, the conversation in front of a blazing log fire turned briefly to the scientific matters we had discussed at Clarkson Close. What would be an acceptable alternative to ice and iron for the composition of cosmic dust? Mesmerised by the flames rising above the burning and sputtering logs, I remember posing a question without too much thought. Could the dust in space be carbon like the soot that lofted up from the fire? Fred's eyes appeared to light up, with excitement or disbelief I could not say. Carbon in the cosmos was after all Fred's baby. It was his prediction of the 7.65 MeV resonance in the nucleus of carbon-12 that opened up new vistas of astronomical thought. Fred, together with Willy Fowler and Geoff and Margaret Burbidge told the world how carbon was synthesised in stars and expelled into interstellar space by supernovae.

But carbon in interstellar dust did not make an immediate impression on Fred, no more than iron did. We all had become used to thinking, following the trend set by Dutch astronomers of the 1940's, that the bulk of interstellar dust had to be formed *in situ* in the gigantic gas clouds of space. And of course oxygen is more abundant there than carbon by a factor of 1.6, so most of the carbon would tend to be tied up in the strongly bound molecule CO. I would guess that these would have been the thoughts that raced through Fred's head at the suggestion of a possible carbon composition of dust.

The following morning after breakfast I joined Fred on his habitual trek over the hills, heading I was told to Bowfell. He had decided that the more ambitious climb to Scafel Pike was not for a first time hiker like myself. Although I was armed with a pair of mountaineering boots and an appropriately warm anorak and headgear, I felt hopelessly ill-equipped for the task ahead. Fred had a battered-looking rucksack on his back in which he carried sandwiches and bars of chocolate supplied by Mrs Cross. I remember Fred saying that a single bar of chocolate would have enough calories to make good the energy we would expend all the way to Bowfell Pike and back. We set out under an azure blue sky almost cloudless save for a stray cumulous cloud or two drifting lazily overhead. The walk began as a more or less gentle amble along the valley which suited me fine, but before long the proper climb began and I was finding it increasingly difficult to keep up with a seasoned hiker like Fred. I believe that Fred noticed my predicament and accordingly modified the route to take us only so far as a lesser peak than Bowfell on this first day.

The gentler walk was welcome, if only to enjoy my first experience of the exhilarating scenery of the Lake District. The landscapes that I had dimly apprehended, thousands of miles away, via the works of the romantic poets, now revealed themselves in all their sheer physical glory. Walking here in the company of Fred Hoyle had a particular poignancy. I recalled Wordsworth's Preface to his *Lyrical Ballads* in which he compares the aspirations of the poet and the scientist:

> "*The knowledge of both the Poet and the Man of Science is pleasure; but the knowledge of the one cleaves to us as a necessary part of our existence, our natural and unalienable inheritance; the other is a personal and individual acquisition, slow to come to us, and by no habitual or direct sympathy connecting us with our fellow beings .... Poetry is the first and last of all knowledge — it is immortal as the heart of man ...*".

Quite early in my association with Fred I could see how the poet and scientist could come together in a single individual.

Some three hours later, we stopped for a spot of lunch. As every mountaineer will tell you the Lakeland weather is extremely fickle, changing suddenly without any warning. So it seemed to be today. The sun had slipped behind an ominously dark cloud and a shower of rain might have been imminent. We found a flat crag to sit upon and as Fred unpacked our sandwiches from his rucksack he glanced thoughtfully at the grey sky. When I casually enquired, "Is it going to rain?" I could not have predicted that his answer would turn out to be a defining moment in my scientific career.

"Not necessarily", said Fred, "These clouds could be saturated with water vapour, but for rain to fall, condensation nuclei are required. These could be charged molecular fragments (ions) or fine dust, but such condensation nuclei need to be present before rain could form." With a moments further reflection he added, "Some have argued that meteor dust could supply nuclei for rain."

From my long experience of Fred I have since realised that a thought such as this rarely remained in isolation in Fred's head: he would begin to make connections with the widest range of problems. With a little prompting from me the connection with dust in interstellar clouds soon came to the fore. If it is difficult to form water droplets in the densities that prevail in the terrestrial atmosphere, how could ice particles condense in the exceedingly tenuous clouds of interstellar space, where hydrogen densities are in the range 10–100 atoms per cubic centimetre? Was the nucleation problem really solved for the case of ice grains in interstellar space? These were questions we pursued in the remaining few evenings by the fireside in the lounge of the Old Dungeon Ghyll Hotel. Fred's writing pad and pen were used for extensive scribbling and we decided that the problem could not really have been solved by the Dutch astronomers. It seemed incumbent on us, therefore, to find a denser place than the interstellar medium to resolve the nucleation problem of interstellar grains.

Chapter 4

Betwixt the Stars

Little did I know as I travelled back with Fred to Cambridge that our walks in the hills of Cumbria in the autumn of 1961 marked the beginnings of a line of research that would, decades later, lead to a new theory of the origins of life. The "Black Cloud" of Fred's novel seemed destined to spring into life.

The black clouds of deep space had, however, to be set in the context of what else there was in the Universe. They show up conspicuously as complex shapes and structures against the more or less uniform distribution of stars in the Milky Way. The Milky Way itself, our Galaxy, is a hundred thousand light year-wide collection of a few hundred billion stars, each one more or less similar to the sun. And our galaxy is one of many billions of similar galaxies that populate the observable Universe.

Black clouds — interstellar clouds — are by no means restricted to our own galaxy. External galaxies often show conspicuous dust lanes, a striking example of which is seen in the case of the Sombrero Hat galaxy NGC4594, which is a galaxy very similar to our own viewed edge-on. The dark lane across the middle represents an overlapping complex of interstellar clouds that collectively obscures the light of background stars. The interstellar clouds in NGC4594, as in our own galaxy, are the birthplace of new stars and planets. So they must clearly have a very special importance in the scheme of things. Depending on where the clouds lie in relation to stars and how dense they are, they take on a wide range of temperatures and properties.

Interstellar clouds are on the average about 10 light years across, and the typical separation between neighbouring clouds is about 300 light years. It should be noted, however, that there is a fairly wide spread in the sizes of these clouds and in their separations one from another. Some are more compact and uniform in their disposition, whilst others are extended and irregular. The more extended clouds appear as giant complexes, showing a great deal of fine-scale structure as cloudlets and filaments. These so-called "giant molecular clouds" are often associated with the formation of new stars.

An interstellar cloud may contain anywhere from ten to many millions of atoms per cubic centimetre. Even the higher values in this density range are considerably lower than the densities that can be attained in laboratory vacuum systems. So it should be remembered that all our intuitive ideas of how gases behave under normal conditions could prove wide off the mark when it comes to understanding what happens under the rarefied conditions of space.

Hydrogen makes up the overwhelming bulk of material in interstellar clouds and occurs in one of three forms: neutral atomic hydrogen (intact atoms with no electrons lost), ionized hydrogen (atoms stripped of their outer electrons), and molecular hydrogen (atoms paired in molecular form as H_2). Molecular hydrogen was first detected using ultraviolet spectroscopy in the late 1960's after my own researches into interstellar matter had begun, although its existence was predicted by Fred Hoyle in the 1950's and indeed exploited in the science fiction novel *The Black Cloud*. Hydrogen molecules are to be found mostly in the denser interstellar clouds that are able to screen off the ultraviolet starlight that could destroy them. A large fraction of all the hydrogen in the galaxy is found to be in molecular form, H_2, and the total mass of the hydrogen is billions of times the mass of the sun.

But what else exists in interstellar clouds besides hydrogen? Information derived from several sources, including solar and stellar spectroscopy and the direct examination of meteorites (rocks of extraterrestrial origin) all have a bearing on the overall composition of interstellar material. Next to hydrogen in order of abundances comes the element helium, which accounts for close to a quarter of the total mass of interstellar matter, although from our point of view this element is inert and uninteresting. Then comes the group of chemical elements carbon, nitrogen and oxygen that together make up several percent of the mass of all the interstellar matter. It is these elements that are of course crucial for life. Indeed, life depends for its function on the unique range of properties of the carbon atom including its high levels of chemical reactivity and its ability to combine into many millions of interesting carbon-based compounds. Next in line are the elements magnesium, silicon, iron and aluminium, which again account for a percent or so of the total interstellar mass. Then a group including calcium, sodium, potassium, phosphorus is followed by a host of other less abundant atomic species. All these chemical elements are synthesised from hydrogen in the deep interiors of stars in the manner worked out by Fred Hoyle and his colleagues in the 1950's. The synthesised elements are injected into interstellar space through a variety of processes, including mass flows from the surfaces of stars. In the case of the most massive stars, the end product of their evolution is a supernova (an exploding star), and it is through supernova explosions that life-forming chemical elements are injected into the interstellar clouds.

The discovery of interstellar molecules (assemblages of atoms) by methods of radioastronomy and millimetre wave astronomy got properly under way a full decade after my journey with Fred had begun. Next to molecular hydrogen the second most abundant and widespread molecule in space turns out to be carbon monoxide. A significant fraction of all the interstellar carbon in our own galaxy, as well as in external galaxies, seems to be tied up in the form of this molecule. Next in line of importance comes the all pervasive molecule formaldehyde, H_2CO, which is present in gaseous form, both in clouds of high density as well as in clouds of relatively low density. In the denser interstellar clouds, particularly in clouds associated with newborn stars, vast amounts of water in gaseous form are found. Water is an important molecule for life, and its close association with newly formed stars and planetary systems would have a vital relevance to our story.

The spatial distribution of interstellar molecules in the galaxy shows wide variations depending on physical conditions such as ambient temperature and density as well as the proximity of clouds to hot stars. As a rule, denser, cooler clouds contain the larger and more complex molecules, whereas lower density clouds and those nearer to hot stars have simpler molecular structures. A region that is particularly rich in organic molecules (molecules that could be connected with life) is the complex of dust clouds in the constellation of Sagittarius, located near the centre of the galaxy. It is in this region that the first tentative detection of an interstellar amino acid, glycine (a component of proteins) was reported, as was the molecule of vinegar and a sugar glycolaldehyde. All these molecules are detected by tuning radio receivers to precisely the frequencies at which the molecules absorb or emit radiation. An inherent difficulty of this technique lies in the need to determine beforehand the correct radio frequencies that characterise particular molecules. Once these are known, the technical problem of finding the molecules is relatively simple. But the actual number of detections could represent a mere tip of an iceberg: a vast number of more complex molecules would inevitably remain undiscovered.

Interstellar organic molecules have been detected by other techniques besides radio astronomy, notably using the methods of infrared astronomy, of which I shall have more to say in future chapters. An important class of organic molecule that is found to be present in quantity are the polyaromatic hydrocarbons or PAH's. These molecules (just as CO) are well-known by-products of the combustion of fossil fuels as occurs, for instance, in automobile engines. They are largely responsible for the suffocating smog that pollutes our larger cities. We shall argue later that even in interstellar space such molecules are most likely to have a direct biological connotation, possibly representing the break-up or degradation products of biological material.

The existence of complex organic molecules in interstellar clouds poses one of the most challenging problems for modern astronomy. The conventional viewpoint is that complex molecules form out of atoms and simple molecules through reactions that take place in the gaseous phase. But because the gas densities in space are so exceed-ingly low (far lower than that present in a laboratory vacuum, as we have said) the reactions occurring between interstellar gas molecules would be too slow to produce any appreciable quantities of the most complex organic molecules. Their presence would have to be explained in some other way.

Comparatively high densities of organic molecules tend to be asso-ciated with regions of the galaxy where new stars (and presumably comets and planets) are forming at a rapid rate. The Orion nebula is a spectacular example of such a region where young stars are evi-dent in large numbers, some even with discs of newly formed plan-etary material around them. Large quantities of organic molecules are associated with the denser parts of the Orion cloud complex, and it would be tempting to link the formation of such molecules with the formation of stars, planetary systems, and perhaps life itself. But more of this would emerge at a later stage of our journey.

In addition to atoms and molecules, interstellar clouds contain an all-pervasive and enigmatic dust component to which I have already referred. Astronomy has struggled to understand the precise nature of this cosmic dust and to discover the circumstances under which such particles are formed. In the autumn of 1961 when I started reading on these matters in earnest I discovered that it was almost an article of faith amongst astronomers that interstellar dust grains were comprised of dirty ice material — frozen water with perhaps a sprinkling of other ices — ammonia-ice, methane-ice and a trace quantity of metals. Furthermore, the firmly held belief was that these particles had to condense, more or less continuously from the gaseous atoms and molecules that were present in the interstellar clouds.

The basic ground rules in this area of astronomy had been laid down in two classic papers and a PhD thesis written in the mid-1940's by the Dutch astronomer H.C. van de Hulst. My researches had led me to a large body of work in an area of physics known as "homogeneous nucleation theory" that was immediately applicable to the interstellar problem. It was easy enough to check the claims of van de Hulst and his colleagues against straightforward predic-tions of the theory. I soon discovered that van de Hulst's ideas were

flawed in several important respects. I was quickly able to verify Fred
Hoyle's conjecture about the nucleation problem — the insuperable
difficulty of forming dust in the exceedingly tenuous gas clouds of
interstellar space. The rain cloud analogy I was introduced to in the
Lake District applied here, only more dramatically. When I commu-
nicated my findings to Fred it was evident that he was pleased. His
immediate response was that I should see van de Hulst in Leiden and
confront him with all the technical details and the difficulties as we
perceived them.

So in November 1961, I made the journey by boat from Harwich
to Hook in Holland and thence to Leiden. I was charmed by
Leiden which was not unlike Cambridge in many ways. It is an historic University town, somewhat younger than Cambridge — the
University being founded in the 16th century — but with a very similar ambience. There were as many bicycles in Leiden as in Cambridge,
but unlike Cambridge's single waterway the River Cam, Leiden was
criss-crossed with boat-lined canals. The birthplace of Rembrandt,
the city is saturated with art galleries and museums.

But the purpose of my visit was to tackle the great man H.C. van
de Hulst. This too turned out to memorable, albeit in a somewhat
negative way. As affable as he was it was clear at the outset that
he was not inclined to engage in any sensible scientific dialogue with
me. I sensed that he was possibly affronted by my challenge of his
views. Only much later did I come to appreciate that Professors in
Europe are traditionally placed on a high pedestal, their authority
rarely being questioned by younger members of staff let alone students. Be that as it may, van de Hulst in 1961 seemed unable or
disinclined to offer a defence of his ideas of grain nucleation that he
had pioneered a decade earlier. Perhaps he had lost interest in the
problem, having moved on to new areas of research. He promptly
directed me to his colleague, Professor J. Mayo Greenberg, at the
Rensselaer Polytechnic in Troy New York, who had now taken up
the cudgels of defending the old ice grain theory against its critics.

My encounters with Mayo Greenberg were still months away.
To conclude the present chapter I shall summarise the first major
objection to the ice-grain model that surfaced as soon as I began

to think seriously about it. True enough, water molecules, if they already exist in interstellar space, could condense on pre-existing particles of dust. But if there is no supply of such dust particles to serve ice grain nuclei, then no condensation could occur. Thus the requirement for any ice-condensation process in space is that condensation nuclei must be injected into the interstellar clouds at a steady rate. In view of the very low gas densities that are present, these nuclei could not be generated at an adequate rate from within the interstellar medium itself. The problem is akin to that of the seeding of water-vapour clouds in the terrestrial atmosphere. One could have highly supersaturated clouds in the atmosphere, but no rain will fall unless condensation nuclei are somehow supplied.

Chapter 5

The Route to Carbon Dust

Faced with the almost insurmountable difficulty of perceiving how ice grains form in interstellar space we took a different approach to the problem of dust formation in 1962. What if the dust was not made of water ice but of carbon, similar to the particles of soot that were rising into the chimney from the log fire we watched at the Dungeon Ghyll Hotel? In this case the formation of carbon dust could occur at much higher temperatures, perhaps, for instance, in the outer atmospheres or envelopes of some cool stars? Such dust grains would also have the advantage of being able to survive in interstellar regions of much higher temperatures.

But what do the astronomical observations tell us about the properties of interstellar dust? What are their precise optical characteristics? How do they behave in relation to the scattering and absorption of starlight? Dust shows up as conspicuous patches of obscuration against the background of distant stars. But several more precise quantitative statements about the nature of the dust were already possible to make in the 1960's. The earliest quantitative investigations were mainly restricted to the way the dust dims and reddens starlight, just as a street light is dimmed in a fog. The first attempts to obtain measurements of interstellar dimming — or extinction, as it is called — of starlight were made in the 1930's. It was found that at a single wavelength (colour) close to 4500 Å the dimming of starlight amounted to a reduction of intensity by a factor of about 2 for every 3000 light years of passage through interstellar space. From this one piece of information alone it was easy to infer that interstellar dimming could only be reasonably attributed to solid particles that have dimensions comparable to the wavelength of light. Much smaller or much larger particles would have to be present with implausibly large densities if they were to produce the observed amount of dimming. And so the dust grains in space were just about as efficient as they could be in blocking the light of distant stars.

With the advent of new techniques in observational astronomy it became possible to measure accurately how interstellar dimming varies with the wavelength of light. This is done by comparing the spectra of two stars which are intrinsically similar, one of which is more dimmed by interstellar dust than the other. This is analogous to comparing two cosmic street lamps, one nearby and another dimly seen through a fog. Such comparisons provide information on the wavelength dependence of extinction caused by interstellar dust. The relationship between extinction and the wavelength of light — what astronomers call the extinction curve — provides an important item of information which has a bearing on the properties of interstellar dust grains. In 1961, this extinction curve was known only over the wavelength interval from about 9000 Å in the near infrared to about 3300 Å in the near ultraviolet. Over much of this wavelength range it was known that the opacity of interstellar dust was proportional to the inverse of the wavelength — in other words, when the wavelength is doubled, the opacity was approximately halved. And most remarkably it turns out that precisely the same type of relationship, the same extinction curve, was found to hold over wide areas of the sky. The requirement for this is that the dust with almost identical sizes and properties exists throughout large volumes of galactic space. In 1961, the best interstellar extinction data over the range of wavelengths 3300–9000 Å was obtained by Kashi Nandy at the Royal Observatory Edinburgh, and this is shown by the points plotted in the upper panel of Fig. 1.

When a set of data points as in Fig. 1 are given, the process of interpreting it involves the construction of a "model" or "hypothesis". In this case a "model" would constitute an informed guess as to what type of interstellar dust could give rise exactly to the relationship between extinction and wavelength as expressed in Fig. 1.

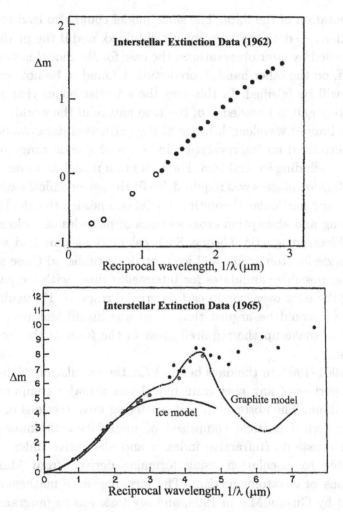

Fig. 1 The manner in which starlight is dimmed by interstellar dust. The points are observations of interstellar extinction: upper frame — as it was known in 1961, lower frame — as known in 1965, with extensions into the ultraviolet. The curves are calculations for particles made of ice and graphite.

A model that one might consider then leads to a calculation of an extinction versus wavelength curve that may or may not match up with the data. If it does match the data, the model can be regarded as being consistent with the observations and would be a valid representation of the data. The same model could also lead to other predictions — data that are not yet obtained, and if the predictions are satisfied by later observations, the case for the model is strengthened. If, on the other hand, a prediction is found to be not true, the model will be falsified. In this way the scientist hopes that he will ultimately gain a knowledge of the true nature of the world.

The limited wavelength range of the extinction data available in 1961 permitted an interpretation in terms of a wide range of dust models, including ice and iron. For each grain model, however, a narrow definition of sizes was required. To fit the astronomical extinction data to any particular theoretical model one needed to calculate the scattering and absorption cross-sections of particles of various radii using Electromagnetic Theory. Such calculations that had already been made for iron grains and ice particles established these models as being possible candidates for interstellar dust, with ice particles having the edge over iron in certain crucial respects. Particularly so because it could be argued that there was insufficient iron in the galaxy to make up the required mass in the form of the interstellar dust.

In 1961–1962, in the days before PCs, the calculation of the optical properties of any new grain model was a major computational undertaking. The absorption and scattering cross-sections of a uniform spherical particle comprised of materials with known bulk optical constants (refractive index n and absorptive index k) was amenable to calculation using formulae derived from Maxwell's equations of electromagnetism. The formulae were mathematically derived by Gustav Mie in 1908, and my task was to program these so-called Mie formulae for use on a high speed electronic computer. This is what I set out to do in the winter of 1961 for a preliminary exploration of our carbon dust models. The optical constants of bulk carbon (graphite) were available at the time only for a few wavelengths, but extrapolations were possible to include the whole wavelength range for which observational data was available. I devised a computer programme in the then-popular language FORTRAN to carry out the calculation, and ran this on the EDSAC2 computer of Cambridge University.

Computers in those days were large clumsy devices. Transistors and printed circuits had not yet been invented and the use of vast numbers of thermionic valves meant that computers were not only large and sprawling, they had also to be cooled efficiently with stacks of air-conditioning fans. This, together with the constant clattering of card readers and output devices, made the environment of the Computer Centre a very noisy place indeed. EDSAC2 occupied several large rooms in the Computer Centre and had less computing power than that of the average laptop that is currently available in 2012. Although my problem was by no means a very difficult one, each computing run on EDSAC2 had to be booked in advance, so developing a new program, particularly debugging it was a tedious affair. After numerous trips to the computer centre in December 1961 and January 1962, I made a remarkable discovery. I found that as long as the diameter of carbon particles were less than about a tenth of a micrometer ($0.1\,\mu$m), the predicted extinction curve was almost indistinguishable from the interstellar extinction observations.

Fred was particularly pleased with this result because it possibly freed the model of any parameter dependence. An identical result was obtained as long as the diameter was less than $0.1\,\mu$m. For ice grains a very precise distribution of sizes was required in order to match the same observations. In view of the invariance of the observed interstellar extinction curve from one direction to another this constraint on the sizes of ice grains was hard to understand and appeared to be a deficiency of the model. The interstellar medium is extremely inhomogeneous with regard to density and one would expect ice particles (if they grew in space) to have wide fluctuations in size from one region to the next. This being not the case we were able to argue that carbon grains had a distinct edge over ice grains in explaining the observed behaviour of interstellar extinction, at any rate over the waveband over which observations were available in 1961. It was possible to calculate how much material in the form of carbon grains we required, and the answer turned out to be one to two percent of the total mass of the interstellar clouds. That is to say an equivalent of one to two percent of the mass of interstellar hydrogen had to be in the form of carbon according to our model.

This was consistent with the availability of carbon in interstellar space.

Carbon in the form of graphite is a highly stable, highly refractory material with a sublimation temperature in excess of 2500 K. Searching for places where carbon grains might form, Fred suggested we might turn to cool giant stars where the surface temperatures were generally below 3000 K. But the major class of oxygen rich giant stars, known as the Mira variables, were not appropriate. With an oxygen excess over carbon in the Mira stars, most of the atmospheric carbon will be in the form of the strongly bound molecule CO and there will be none left for graphite formation. There was however a class of red giant stars known as the carbon stars, the so-called R and N variable stars. The N-stars have surface temperatures that varied cyclically between 1800 K and 2500 K over a period of about a year, and they are known to have more atmospheric carbon than oxygen. Thus although oxygen would again form CO, there would be an excess of C that is able to condense into solid particles when the temperature fell below some critical value. Again a computer programme had to be deployed to determine the physical state of the excess carbon as the temperature varied between 1800 K and 2500 K. Using standard theories of nucleation in a homogeneous saturated gas, we showed that carbon particles were able to nucleate and grow in the stellar atmosphere as soon as the temperature fell towards 2000 K. We further showed that graphite particles of radii of a few hundred Angstroms would grow from initial nuclei and be expelled into interstellar space, the expulsion being caused by the pressure of light from the parent star. There is observational evidence which points strongly to the existence of dust around carbon stars. The variable and highly luminous carbon star R Corona Borealis is a spectacular example. Here we see direct evidence of a star erratically puffing out clouds of carbon soot into the interstellar medium. More recently astronomers have also detected thermal infrared radiation from carbon stars which is consistent with the presence of heated graphite particles.

Throughout much of 1962, Fred was busy with Willy Fowler's visit to Cambridge and their mammoth project on nucleosynthesis.

This did not, however, deter him from an active engagement in our own work on carbon grains as the story began to unfold. It did not take more than five months after returning from the Lake District for us to produce a paper entitled "On graphite particles as interstellar grains" for the Monthly Notices of the Royal Astronomical Society. This was submitted at the end of May 1962 and our paper appeared later that year (F. Hoyle and N.C. Wickramasinghe, *Mon. not. R. Astron. Soc.* **124**, 417, 1962). This publication represented our first step in the direction of cosmic biology although we did not recognise it as such at the time. I soon followed up our first paper with a series of others in Monthly Notices describing detailed calculations of the properties of carbon particles, including the possibility of ice mantles condensing around them in interstellar clouds. I also made a prediction of the properties of interstellar grains based on new measurements of the optical constants of graphite, which were brought to my attention by the Belgian astronomer C. Guillaume. The prediction was that if the extinction curve was extended into the ultraviolet, a strong absorption feature centred at 2200 Å would be seen (N.C. Wickramasinghe and C. Guillaume, *Nature* **207**, 366, 1965).

Throughout most of 1963, my interaction with Fred was confined mostly to reporting progress on new developments of the carbon grain story. This was my opportunity to establish myself as an independent researcher which I did with much enthusiasm. Early in 1963, I began writing my PhD thesis and thinking seriously about what I might do when my Commonwealth Scholarship came to an end in September 1963. Our work on the carbon grains and on grain formation was opening up new vistas of research in this field, so I did not relish the prospect of returning to a University position in Ceylon, where I may not have been able to pursue my research.

On Fred's advice I decided to apply for College Fellowships in Cambridge. The competition for such fellowships is notoriously stiff. Competitors include young researchers from all academic disciplines, so the best in one field had to be compared with the best in others, a very difficult task for electors, as I was to later discover. So I was delighted when Jesus College elected me to their Research Fellowship in 1963. This opened a new chapter in my life as a research scientist and a Cambridge don. The Fellowship included a modest stipend, free rooms and meals in College. I moved into my rooms in a new building in North Court to finish my PhD thesis whilst also taking on a modest amount of undergraduate teaching for the College.

Compared to my somewhat isolated life as a graduate student at Trinity, I began to enjoy the fellowship of my colleagues at Jesus who came from a wide range of academic disciplines. The President of Jesus College at the time was a mathematician Alan Pars, who had taught Fred Hoyle as well as my father in the 1930's. Alan Pars very generously took me under his wing and made me feel comfortable in my new surroundings. Jesus College had no great traditions in astronomy at the time, but had in its history an eminent alumni: John Flamsteed (1646–1719), the original Astronomer Royal of England. Flamsteed's monumental publication of the first extensive star catalogue set the highest standards for observational astronomers who were to follow him. His work was also a great boon to navigation, providing a basis for the accurate determination of position at sea. In the congenial setting of my rooms at Jesus I was able to pursue my researches on various aspects of interstellar grains. It seemed clear to me that a major paradigm change was round the corner — a shift from volatile ice grains to refractory grains, grains that had to be based largely on the element carbon. It was also a shift from ideas of grain formation in diffuse interstellar clouds to condensation of dust in much denser stellar environments. The transition, however, was not as smooth and painless as I had anticipated it to be.

The publication of our papers on interstellar graphite led to a head-on clash with proponents of the ice grain theory in the USA, particularly J. Mayo Greenberg and his band of collaborators. New techniques in observational astronomy were now making it possible to study the behaviour of interstellar dust outside the visible wavelength range — at longer wavelengths beyond red (the infrared) and shorter wavelengths beyond blue (the ultraviolet). Observations in the infrared waveband were soon used for discriminating between grain models. R.E Danielson, N.J. Woolf and J.E. Gaustad were the first to search for a characteristic absorption feature of water ice at the infrared wavelength of 3.1 μm in the spectra of highly dimmed stars. By 1965, the lack of a 3.1 μm water ice band in the spectra of several stars led to the conclusion that ice particles if they exist at all can make at most a very minor contribution to the interstellar dust. Although 3.1 μm bands were later detected in some astronomical sources, these most probably arose from matter local to the sources themselves, from dense circumstellar clouds not from dust in the general interstellar medium.

Spectroscopic data in the ultraviolet region of the electromagnetic spectrum at wavelengths shorter than 3000 Å were first obtained by T.P. Stecher, R.C. Bless and A.D. Code using equipment carried on rockets and satellites. The most conspicuous feature in the ultraviolet interstellar extinction curve observed by T.P. Stecher was a broad hump centred on the wavelength 2175 Å. This was exactly as we calculated for spherical graphite particles of radius 0.02 μm, as shown in the comparison of the lower panel of Fig. 1. The ice grain model could not produce such an ultraviolet feature and must therefore be deemed to be inconsistent with observations.

The new infrared and ultraviolet data provided enough reason for a conference on interstellar grains to be convened. A conference was promptly organised by J. Mayo Greenberg, held at Rensselaer Polytechnic Institute in Troy New York from 24 to 26 August 1965.

This was my first visit to the United States and memories connected with it still linger in my thoughts. Landing at La Guardia Airport in New York I spent a day walking around in the shadow of skyscrapers in Time Square. But after the initial bedazzlement the stark reality of American society begins to impinge. Concepts of aesthetics and grandeur differ markedly from those of the old world.

The next day I arrived at Troy for my first scientific conference. I felt a heavy burden of responsibility to defend the graphite grain theory that Fred and I had proposed. I had tried to persuade Fred to join me on this trip, but it turned out to be impossible for him. Greenberg, the local host of the conference, and the person van de Hulst had referred to, was our sworn enemy. He was determined to defend the ice grain model using every device that was available to him. His most gallant attempts at producing 2200 Å extinction features using "trimodal size distributions of infinite cylinders made of ice", were more ridiculed than lauded at this meeting. Other papers presented here included one by Bertram Donn and Ted Stecher from NASA's Goddard Space Flight Centre on a graphite model, which was not dissimilar to ours. Perhaps the most remarkable paper of the entire meeting was one by Fred M. Johnson in which he argued that many unidentified diffuse absorption bands in stellar spectra at visual wavelengths can be caused by a derivative of chlorophyll (a porphyrin, the green colouring substance in plants) — a paper that was several decades ahead of its time.

The period between 1965 and 1967 was particularly difficult for Fred. His negotiations with Cambridge University and the Government for starting an Institute of Theoretical Astronomy had suffered many time-consuming setbacks. There was also an important development in cosmology which was quickly seized upon by his adversaries. In 1965, Arno Penzias and Robert Wilson accidentally discovered a diffuse background of microwave radiation emanating uniformly from all directions in the sky with a temperature of 2.7 K. Excluding all other possible causes of this radiation they interpreted this signal as the relic radiation from a Big Bang Universe. This so-called cosmic microwave radiation was described as the last nail in the coffin of the Steady State Universe. With the enormous weight of the propaganda that was mounted it appeared to me that Fred began to accept defeat for a while. But that did not last too long.

Chapter 6

A Theory Takes Shape

After two years of working as a postdoctoral researcher on interstellar grains the time had come to take stock of what had been achieved. When I first started my studies in 1960, interstellar dust was considered to be a barren field for research. To most astronomers the presence of dust was a nuisance, its only effect being to hinder the observations of distant stars. All they needed were simple rules to correct measured intensities of starlight to compensate for the presence of dust; beyond that interest in dust was minimal. In the few years of my research this situation seemed to be changing. Interstellar grains were certainly coming into vogue with new observational opportunities and techniques paving the way to new ideas, and to ambitious programmes of work. It cannot be denied the entry of Fred Hoyle and myself into this field had a part to play in this transformation of attitude.

The group of international astronomers who attended the conference in August 1965 were aware of my work on graphite grains and new avenues became open. Many were sympathetic to the idea of carbon grains, evincing more than a passing interest to challenge the ailing ice grain paradigm. Bertram Donn of NASA's Goddard Space Flight Center was particularly keen in pursuing further the nucleation aspects of the carbon dust theory, to understand how a gas of carbon atoms could form condensation nuclei and thence grains. Some years earlier, he and John R. Platt had argued the case for large unsaturated organic molecules nucleating in interstellar space. He now felt that a similar process could more easily operate in carbon stars. In the so-called "Platt particles", the extinction of starlight (absorption and re-emission) occurs due to electrons jumping between discrete energy levels. Although I had pointed out to Donn that it was difficult to explain the details of the observed interstellar extinction by this process, the formation of carbonaceous particles certainly merited careful study. The work of Platt and Donn in 1960 gave what was perhaps the first hint of large organic molecules occurring in space. It was therefore not surprising that Donn thought it worthwhile to explore further aspects of our ideas on the theory of carbon grains. He accordingly arranged with the Department of Physics and Astronomy at the University of Maryland for me to be appointed a Visiting Professor in the Summer and Fall of 1966. In this way I would be able to interact with his Astrochemistry research group at the Goddard Space Flight Center in Greenbelt which was only a few miles from College Park along the freeway.

But long before my meeting with Donn, Fred had arranged, through his friend Willy Fowler, that I spent the fall semester of 1965 at the Kellogg Radiation Laboratory at Caltech in Pasadena. To accommodate all these arrangements I obtained leave of absence from Jesus College for the entire academic year 1965/66 to work part of the time in the United States and the rest in Sri Lanka. The Sri Lankan stint was arranged in order to take up an appointment as a Visiting Professor of Mathematics in a newly-formed Vidyodaya University in Kelaniya. Here I thought I might be able to assess a possible long-term option of returning eventually to Sri Lanka. The research task that I set myself to do during the Sri Lankan spell was to complete a technical monograph on "Interstellar Grains" that was commissioned by Chapman and Hall (London) in their International Astrophysics Series. This, I felt, was an important step that would put our new ideas on grains firmly on the scientific map.

The first phase of my plan for 1965/66 did not turn out to be as successful as I had hoped it would be. Indeed my semester at the Kellogg Radiation Laboratory in Caltech took off to a bad start, when on my first evening I decided to take a stroll along the wooded streets just outside the campus precincts. Within minutes a police car pulled up beside me and a slow-witted, thick-set cop jumped out to interrogate me as to my business of walking on the street. Everyone, he explained, went by car and walking here was simply against the law. My crime, for which I was let off with a warning, was compounded by the fact that I did not carry any money or form of identity. This incident, marking an infringement of my civil liberties, was an unfortunate introduction to life in the United States.

I had expected to be greatly stimulated by the scientific atmosphere of Caltech, one of the most distinguished centres of scientific research in the world. But this was not to be. During my first weeks at Kellogg all my attempts to interest astronomers with problems connected with grains did not meet with too much success, and this was a disappointment to some extent. Most astronomers here were preoccupied with much bigger problems of cosmology and nucleosynthesis. The recent discovery of the cosmic microwave background gave an added impetus for such a narrow focus. This lack of enthusiasm for problems connected with interstellar grains might sound strange from the perspective of 35 years on; Caltech was later to take a leading role in the development of infrared astronomy, which of course had a direct bearing on the nature of interstellar grains. Infrared astronomy reached maturity around 1968, so my visit in 1965 was perhaps only a few years too early. I left Caltech at the end of November 1965, a little sooner than I had anticipated, and headed home to Colombo, via Hawaii and Tokyo.

Landing at Katunayake Airport with the prospect of a full five-month stretch in my homeland inspired me with joy. Compared with the maze of freeways around Los Angeles that I had just left, the journey from the airport to my home in Colombo appeared somewhat medieval. Pedestrians, cyclists, bullock carts, cars and buses jostled for space along a narrow winding road. Disorderly humanity wound its own path along crowded routes and a twenty-five mile journey took well over an hour and a half.

In a span of five years the city of Colombo as well as the immediate environs of my family home had changed. The capital had become more congested and polluted, and one sad consequence for me was that the magnificent spectacle of the night sky, that I had so often enjoyed from my front garden, had now become a rarity. Street lights just outside our house and a more general haze of light pollution had spoilt forever our capacity to enjoy the beauty of the cosmos.

The experience of being installed as a Professor in a new Sri Lankan University turned out to be more daunting than I imagined. I found myself being instantly drawn into a great deal of arduous administration. This was perhaps to be expected in a new university institution, but I did not find this part of my work particularly agreeable or rewarding. A poorly stocked University Library made any chance of continuing front-line research in my highly specialised field somewhat remote. This would of course have been different had the internet come 40 years earlier! Fortunately I had brought with me all the material that I needed for writing my book on Interstellar Grains. This is what I was able to do in the time that was left for me between my other commitments. And for the rest I simply had to bide my time until I returned to Cambridge, back to my Jesus Fellowship and eventually to a staff position at Fred Hoyle's new Institute of Theoretical Astronomy. The Institute itself became a statutory reality in July 1966, with its buildings in Madingley Road to be completed the following year.

By far the most momentous event in my life, one that was to have a profound influence on the course of my journey with Fred, happened during my sojourn in Sri Lanka. This was my meeting with Priya Pereira, a law student at the University of Ceylon, and our subsequent marriage in April 1966. A beautiful, intelligent, talented girl aged 20 enters my life and this story. As she still continues to complain in jest, she was wrenched from her family and a highly promising legal career (her father was a leading lawyer) to become my wife. As our work on the grains turned into directions that became ever more controversial, Priya's steadfast support and encouragement, and most of all her combative temperament was a crucial factor in our progress. It could be said with all honesty that Priya helped me endure the slings and arrows of outrageous fortune, in the instinctive belief that the ideas so carefully thought out by Fred Hoyle and myself must turn out to be right! Little did Priya realise in 1966 that she was stepping in with me into a lion's den of acrimony and controversy in the years that lay ahead.

But the worst of all that was still a full decade into the future. In April 1966, I went on the second sea voyage of my life, this time in the delightful company of Priya, from Colombo to Southampton, bidding farewell to family and friends, and the country of our birth. We arrived together in Cambridge to set up our first house in a flat in Jesus Lane that was provided for us by Jesus College. I had just two months to resettle in Cambridge and introduce Priya to all my friends, particularly the Hoyles.

Then we were on our travels again, this time by ship *SS France* to New York, and then on to College Park, Maryland where we would be based for three months. Although Priya and I were agreed that we would not be living full-time in the States, we made the most of our stay here, spending our weekends touring the national parks of Virginia, Baltimore and Washington in an old Chevrolet that we had bought for $50.

My office on campus was located in the Department of Physics and Astronomy headed at the time by the Dutch radio astronomer Gart Westerhout. My appointment as a Visiting Professor was in the University of Maryland, but since the funding for my post came from NASA, most of my collaborative links were with Bert Donn's Astrochemistry group at the Goddard Space Flight Centre in Greenbelt. It took me a while to get accustomed to the new environment and particularly to the work ethic that prevailed. I had got used to working Cambridge-style on an individualistic basis, even in my collaborations with Hoyle. Here at Goddard I was drawn into more restrictive communal research atmosphere. People liked to talk a great deal and to work, or appear to work, in big teams. There was always an impression of great industry, with large numbers of people working at a frenetic pace on a single problem. But the output was not always commensurate with the manpower or effort that was expended.

As agreed with Bert Donn my task in the three months at Maryland was to investigate further the problem of the formation of graphite grains. The earlier work on this subject that was already published by Fred and myself in Monthly Notices came to be carefully re-examined and elaborated upon by a team of four authors led by Bertram Donn. The work as I remember it involved a tiresome succession of meetings and conferences occupying a great deal of time. And even at the end of my stay a final manuscript for submission was not agreed upon. Many letters and drafts had to be exchanged across the Atlantic, and the ultimate result was the publication two years later of a paper in the *Astrophysical Journal* (B. Donn *et al.*, *Astrophys. J.* **153**, 451, 1968).

American scientists in 1966 were clearly ahead of the British in the practice of counting research papers and citations in a race to justify their existence, and to secure continued public funding for their projects. The levels of stress generated by this process were visibly detrimental to the health and well being of the scientists concerned and also to the progress of science. Nowadays almost every scientist has come to take this practice for granted, with universities and research groups vying to get the maximum number of papers published in the shortest time and in the so-called high impact journals.

Our papers on interstellar grains published throughout the period 1962–1968 and beyond were all in high impact journals, and impact they certainly had in helping to change astronomical fashions from volatile ice grains to refractory grain models. A connection between grains and organic material slowly entered my thinking when I began to look at the detailed thermodynamic equilibrium compositions of the atmospheres of cool stars. In carbon stars, for instance, the equilibrium calculations I performed whilst I was in the States showed clearly the presence of large numbers of hydrocarbon molecules which would inevitably be expelled into interstellar space along with the grains. I had suggested a stellar origin for some of the organic molecules detected in the interstellar clouds, but an explicit link between grains, biochemicals and life had to remain in gestation for some years to come.

Chapter 7

The Institute of Astronomy:
The Vintage Years

In the summer of 1967, the Institute of Astronomy in Cambridge came into physical existence in a well-equipped open plan building in the midst of a meadow off Madingley Road. It seemed to be strategically placed between two friendly institutions — the Cambridge University Observatories on the one side and the Geophysics Department on the other. Fred's cosmological adversaries at the Mullard Radio Astronomy Observatory were located only a couple of miles away on Madingley Road, but as far as interaction was concerned they may as well have been as far away as the Moon. Martin Ryle and his team were continuing in their single-minded pursuit of disproving Steady State Cosmology. From their studies of the counts of radio sources to various intensity levels, they claimed that radio emitting galaxies appeared to be closer together as one goes further back in the Universe, showing that the Universe could not be in a steady state.

The real crisis for Steady State Cosmology was, however, not radio source counts but the new discovery of a cosmic microwave background with a temperature of 2.7 degrees above absolute zero (2.7 K). To offer any credible defence of Steady State Cosmology it was imperative to explain this background by a process that was unconnected with early hot phase of a Big Bang Universe.

After my return from the United States, Fred and I had many discussions exploring possible non-cosmological interpretations of the cosmic microwave background. The question as to whether grains could play a role naturally arose, but I had difficulty in arriving at any tenable model. We even briefly entertained the possibility that an isotropic cloud of dust enveloping the entire solar system could somehow succeed in degrading sunlight energy to a mere 3 degrees! Nothing was working. Fred alerted me to several strange coincidences that may be relevant to this issue. The density of energy of starlight in the galaxy was very close to the density of energy in the cosmic microwave background. Further he pointed out that the energy released in the conversion of hydrogen to helium averaged over a cosmological volume (a large part of the entire observable universe) had a density that was also similar to the energy density of the microwave background. The implication of all this is that a perfectly absorbing and emitting object (a black body) placed either in interstellar space or intergalactic space would take up a temperature close to that of the microwave background, 2.7 K. Could all this be dismissed as a miraculous coincidence? I thought not. But the challenge was to find approximations to black body absorbers and radiators in space. If some mechanism can be found for thermalising the energy of starlight, there would be a hope for explaining the microwave background in terms of ongoing astrophysical processes within a steady state universe. But the situation was by no means easy. If one thought of solid interstellar grains of radii of about a tenth of a micrometer, they would be exceedingly efficient absorbers of visual starlight, but very poor radiators at long wavelengths. The reason is that such a small particle acts as a very inefficient antenna for re-emitting the absorbed starlight at very long wavelengths, that is to say wavelengths that are much greater than the size of the grain itself. Thus a normal interstellar dust grain would heat up due to an internal greenhouse effect, and take up a temperature of 15–30 K, much higher than the blackbody temperature which is nearer to 3 K.

If, however, there are oscillators (radiators) within the grain that can vibrate at microwave frequencies 10–100 Megahertz (MHz) the situation could be different. Grains could then yield high outputs of radiation over a microwave waveband. This is precisely what we discovered in 1967 for the case of impurity atoms that are very weakly bound in solids. Fred Hoyle and I argued in a paper in *Nature* (**214**, 969–971, 1967) that the cosmic microwave background could perhaps include a contribution arising from these types of grains. Jayant Narlikar and I followed this up with two attempts to work out detailed spectra and to estimate isotropy of a background (how uniform the radiation is in all directions in the sky) arising from such a process (*Nature* **216**, 43–44, 1967; *Nature* **217**, 1235–1236, 1968).

Yet another attempt at a non-cosmological explanation of the microwave background came from a collaboration with Vincent Reddish, during one of the many visits to the Royal Observatory in Edinburgh that I made in those days. We discovered from published data relating to solid and liquid hydrogen available at the time that the condensation of solid hydrogen onto grains could occur in interstellar clouds at a temperature close to 3 K — very close indeed to the temperature of the microwave background (N.C. Wickramasinghe and V.C. Reddish, *Nature* **217**, 1235–1236, 1968). For Fred, though, this was an opportunity to revive his enthusiasm for Steady State Cosmology. A very elegant logical argument could be made. If the most abundant chemical element in the Universe, hydrogen, could freeze at 3 K several important consequences would follow. Freezing of hydrogen in dense clouds reduces gas pressure within the clouds, and hence leads to the collapse of clouds. The formation of stars and galaxies that follows tends to increase the energy density of starlight, because they put out more radiation into space. This in turn cuts off the process as soon as the local temperature rose above the freezing point of hydrogen. A feedback loop of this type acting like a thermostat would maintain the background temperature at precisely 3 K. This impressive argument was published in 1968 (F. Hoyle, N.C. Wickramasinghe and V.C. Reddish, *Nature* **218**, 1124–1126, 1968). Unfortunately for us it turned out that the thermochemical data used by us in 1967 was inaccurate and came to be revised. The new data, which was brought to our attention by our old friend J. Mayo Greenberg, showed that the temperature of freezing of hydrogen in interstellar clouds was nearer 2.3 K than 2.7 K. This of course dealt a death blow to the solid hydrogen explanation of the cosmic microwave background. We were back to square one.

But there were more urgent matters that demanded our attention. The Institute was now in full swing and Fred had entrusted me with the job of getting a core library together, a task which took up several weeks of my time. And of course there were other interesting astrophysical problems to turn to. A state of the art IBM Computer up and running in a building adjacent to the main Institute beckoned us to tackle computational problems that were hitherto beyond our reach. I continued my attempts at modelling various aspects of interstellar grains, including their infrared emission and absorption properties, whilst also beginning to think about another interesting computational project. This was the problem of the condensation of the planets.

The problem itself was not new and goes back to the old nebular hypothesis first suggested by Immanuel Kant and formulated in a scientific framework in 1796 by the French mathematician Pierre Simon de Laplace. It is generally believed now that the material of the sun and planets once comprised a single diffuse cloud of interstellar gas and dust. We see similar clouds widely distributed between the stars of our Galaxy, and as we have pointed out earlier it is clear that stars condense from such clouds. The manner in which such a cloud of gas and dust can separate to form the sun and the planets is less clear, however. An early theory suggests that the first step was the formation of a double star system consisting of the sun and a much smaller companion star going round one another; then the smaller star is supposed to have exploded spreading its material in a disc out of which the planets condensed. Another theory states that a single star, the sun, contracted from the cloud leaving behind a debris of gas and dust from which the planets subsequently condensed. Yet another possibility is that the planetary material left the contracting sun during an early stage of its evolution.

Fred and I argued in favour of this last possibility, believing that the others are more or less untenable. The vital clue lies in the very simple fact — that the sun rotates about its axis once in about 26 days. If the sun contracted from a cloud of gas and dust, similar in size and speed to the clouds we see between stars, then it is possible to predict quite unequivocally that the sun must have rotated several hundred times faster than it now does. In fact it should rotate on its axis once in a fraction of a day. This is on account of a physical law which asserts that the rotational momentum of a mass of material, which is not acted on by external forces, cannot change. The consequence must be that the compressed smaller object must spin faster in order to contain all the rotational momentum of the original dispersed cloud. The situation is analogous to ballerina who spins faster as she draws in her arms. What has happened to this rotational momentum of the sun? If one adds up all the rotational momentum of the planets it turns out that here indeed is all the missing rotational momentum of the sun. It appears that, in some way, the contracting sun had given up most of its rotational momentum to the much smaller mass of the planetary material.

There is one very simple manner in which such a transference of rotational momentum can take place. As the cloud of gas and dust which formed the sun contracted, it started off by taking in all the rotational momentum. The cloud would thus rotate faster and faster as the contraction proceeded. A critical stage is reached when the cloud's gravity cannot hold up to the strong rotary forces that begin to develop, particularly near the flattened equatorial regions. The result will be that a disc of material which later formed planets is squeezed out from the equator.

At this stage we can estimate that the young Sun would have been about 10 times larger than at present, and that its surface temperature was about 3000 K, about half its present value. The planetary matter ejected at this time is slightly ionised (that is, separated into nuclei and electrons) and thereby anchored to the parent sun by magnetic fields in a manner discussed by Hannes Alfven. It is this anchoring (acting like spokes in a bicycle wheel) that actually caused the transference of rotational momentum from the sun to the planetary material, resulting in the slowing down of the rotation of the sun. The speed of ejection of the material must have been sufficiently great for the planetary disc to have been able to reach out to the distances of the outermost planets, Uranus and Neptune. It is at such distances that large quantities of the lightest gases, hydrogen and helium, must have escaped into space. Indeed it appears that as much as 6/7 of the total mass of the planetary disc must have been lost to the solar system in this way.

My brief now was to deploy the Institute's computer to work out the chemical history of the planetary material from the stage the disc leaves the Sun's equator in the form of a gas to the time of planet formation. The actual calculation took up about 25 hours of computing time on the Institute's computer, although nowadays on my laptop at home the same calculation would take only a few minutes!

When the planetary material leaves the sun its temperature is about 3000 K. It consists of gaseous atoms and ions in uncombined forms such as Hydrogen, Helium, Oxygen, Carbon, Nitrogen, Silicon, Magnesium, Iron and Sulphur. In the course of its journey outwards, the temperature, density and composition of this material changes in a manner which can be calculated from known physical and chemical theories. Various types of molecular species begin to form as the gas recedes from the Sun and consequently cools off. It is possible to calculate the distances from the Sun at which various types of solid particles and liquid droplets begin to condense out of the gaseous mixture. Our calculations indicated that the first such condensations must take place almost precisely at the distances of the innermost planets — within the orbits of Mercury, Venus, Earth and Mars. The temperature of the gas at this stage is about 1500 K, and we found that fine particles of iron, magnesium oxide and various silicates begin to form — precisely the constituents of the terrestrial planets. At these high temperatures iron particles tend to be very sticky and collisions between such particles would lead to the formation of quite large sized objects measuring 1–10 m across. Whilst a fine smoke of particles would tend to be swept along with the gas, objects of 1–10 m across would fall out of the expanding disc at the distances where they are formed. These metric sized blobs of iron and magnesium silicates then aggregated further and eventually led to the formation of the inner planets, including the Earth.

The most notable success of this theory was that we could give a plausible explanation for why the Earth and inner planets are made up mainly of iron and magnesium silicate in various mineral forms, and why they are found at their present distances from the sun. It seems that these features were determined almost entirely by the density and temperature at the surface of the sun when the planetary material left it, and by the thermochemistry of the various gases involved. Water and carbon dioxide could have been trapped only in rather small quantities at the Earth's distance — probably in the form of hydrated silicates and carbonates. This again is in agreement with respect to the quantities of such substances we find on the Earth.

It is only when the planetary disc expanded out to the present orbital distances of the outer planets Uranus and Neptune that carbon dioxide-ice and water-ice could condense, first into solid particles, then into 1–10 m sized bolides and next into hundreds of billions of cometary-sized icy objects. Collisions between these icy bodies occurring over hundreds of millions of years led to three further effects:

1. Some icy comets were deflected into orbits that made them collide with the Earth and other terrestrial planets, leading in the case of the Earth to the acquisition of volatiles such as water and carbon dioxide. The Earth's oceans and atmosphere came in this way.
2. Some cometary objects were deflected outwards to form a shell of comets around the sun.
3. Approximately half of the icy objects ended up in planetary accumulations we now recognise as the planets Uranus and Neptune.

The attractive feature of this theory is that it is able to account for the observed chemical compositions and distances of the various planets in the solar system in a sensible way.

During the time of the accumulation of the outer planets, the entire solar system remains immersed in the remnants of the molecular cloud from which it formed. In a contact sweeping through the cloud, the shell of comets mops up large quantities of organic molecules and dust of the kind we described in an earlier chapter. It is in this way that comets came to incorporate pristine interstellar material.

Our new theory of planet formation was published in *Nature* in February 1968 (*Nature* **217**, 415–418, 1968). Its impact on ideas relating to interstellar grains and to life in space eventually turned

out to be more far-reaching than we imagined at the time. As a
source of interstellar grains the planet forming processes were just
as important, if not more important, than sources from cool giant
stars. When the inner planets were accumulating not all the iron
and siliceous dust would have coalesced into larger bodies that
remained gravitationally bound to the solar system. A significant
fraction would have been carried outwards and expelled into interstellar space through the action of radiation pressure by an early superluminous sun. This would provide one source of refractory metallic
and siliceous dust in interstellar space. We discuss later how nascent
planetary systems possessing comets like our own solar system could
also provide a source of organic interstellar dust, perhaps even life.

Fred was working at this time with Willy Fowler on nucleosynthesis in exploding supermassive stars in the centres of galaxies. The
question naturally arose whether dust could condense directly from
the heavy element enriched gas that flowed out from such exploding supermassive stars. To tackle this question I had once again to
deploy my chemical equilibrium program. In a calculation very similar to that described earlier for the case of the planetary disc I was
able to show that, with solar abundances at the source, magnesium
oxide, iron and silica particles must again form when the temperature of the expanding supernova shell cooled below about 1200 K.
We thought now only in terms of spherical dust particles for all these
three compositions — including iron. It would later turn out that iron
was much more likely to form as long whiskers, a feature that would
have a relevance to the problem of the microwave background that
was discussed earlier in this chapter.

Chapter 8
Winds of Change

Just as I was beginning to feel settled in our life in the UK, racial tensions began to grow on both sides of the Atlantic. On April 11th 1968, Civil Rights leader Martin Luther King was gunned down in Memphis and a wave of race riots spread across major US cities. Within an amazingly short space of time ripples of racial disquiet reached Britain. On April 21st, Enoch Powell made his historic "rivers of blood" speech. Powell, a distinguished classical scholar, made his point most eloquently: "As I look ahead I am filled with foreboding. Like the Romans I see the River Tiber foaming with much blood". He went on to say that Britain must be, "mad, literally mad as a nation to admit 50,000 dependents of immigrants into the country every year". The present situation he concluded is like a nation "busily engaged in heaping up its own funeral pyre". Conservative Party leader Ted Heath was quick to denounce Powell's speech and expel him from the shadow cabinet, but for those of us who were attempting to adopt Britain as our home, these developments were a source of anxiety and insecurity.

My work was a solace. Fred and I now continued to explore non-carbon based contributions to interstellar dust. Our papers on condensation sequences of solids in the expanding solar nebula and in supermassive stars (*Nature* **217**, 415–418; *Nature* **218**, 1126–1127, 1968) detail our views at the time. We were considering then the possibility of particles comprised of iron and siliceous material coexisting alongside with graphite in interstellar space. However, because the elements silicon and iron were down in abundance from carbon and oxygen by a factor of more than 10, we did not think there was an easy way of detecting such grains from the overall extinction behaviour of interstellar dust.

Whilst this work was in progress, Fred was involved in a government project to build a telescope in collaboration with the Australians. He played a key role in the choice of site for a southern hemisphere observatory, and also in the negotiations for securing UK funding. In fact it was the formidable Margaret Thatcher, then UK Education Secretary, who had to be convinced that Britain should make a financial commitment of this kind. After trying various approaches, Fred informed her that television viewing figures for recently screened programmes connected with astronomy ran into several millions, far more than for any other factual programme. That, according to Fred, instantly won her support.

The summer of 1969 had a special significance for Priya and myself. Fred's wife Barbara visited Sri Lanka with her friend Viv Howes, and we had the pleasure of looking after them. Whilst they were in the capital they stayed in my parents' house in the pleasant residential suburb of Colombo 4, but they spent many days touring the island. We accompanied them on most of their trips and together with our visitors we saw things and places we had never seen before. The 21 years of our lives that we spent in Sri Lanka were evidently not enough to have explored all the remarkable sites of this richest of isles. Our trips took us to the ancient ruins of Anuradhapura and Polonnaruwa, to the hill country capital Kandy, to tea plantations and to the beaches on the West coast of the island. Whilst in the hill-country station of Ella, the place that made the greatest impression on all of us was "Land's End". In the first morning mists at the guest-house where we were staying, we experienced that peculiar sense of the infinite, of our dwelling here on Earth giving way to the sky and the eternal cosmos. Gigantic stone statues of the Buddha set among the arid planes of central Sri Lanka reinforced this feeling. Anita (infinity) lies at the heart of the Buddhist philosophy — a world without end or limit, into which the individual may at last be sublimated.

The year 1969 witnessed two great triumphs of our technological civilization — in air travel and space travel. A long 66 years after the Wright brothers took flight in their primitive pedal aircraft in 1903, Concorde's maiden flight took off. And in the realm of space travel the first human being set foot on the moon. In the small hours of July 21, flickering pictures of Neil Armstrong taking his first steps on the lunar soil beamed back to our television screens and to a world agog with excitement and wonder. The first moon walk was undoubtedly a great technological achievement and a commendable success for NASA. Indeed, it remains Science's most popular gesture to date, creating the appealing impression of Man as the supreme ruler of the Universe. Its relevance to the advancement of astronomical knowledge was, however, a matter of dispute.

Nonetheless, the events of 1969 paved the way to a new era of astronomical observations conducted from space. A new generation of telescopes and instruments carried aboard satellites came into operation. For example, NASA's Orbiting Astronomical Observatory 2 (OAO2) began to yield a wealth of new data on ultraviolet spectra of stars that had a direct bearing on the composition of interstellar dust. The existence of a conspicuous ultraviolet absorption feature centred at about 2175 Å was confirmed, and a continued rise of interstellar extinction further into the ultraviolet was discovered. At this time Fred and I were still pretty much tied to the idea of graphite grains causing the mid-ultraviolet extinction hump, with other grain materials being responsible both for the visual extinction of starlight, and the further rise of extinction into the far ultraviolet. A striking result that emerged from the OAO2 data was the invariable wavelength placement of the centre of the 2175 Å absorption hump from star to star. This, according to our models, demanded the presence of spherical graphite particles with radii almost exactly fixed at 0.02 micrometres. Even as early as 1969, Fred and I were beginning to feel a little uneasy about the artificiality of these assumptions. But a properly formulated alternative to graphite lay a few years into the future.

The period 1967–1970 saw also the emergence of the new discipline of infrared astronomy. The first deployment of liquid nitrogen-cooled detectors operating at 2.2 micrometres at the Mt. Wilson and

Palomar Observatory yielded a catalogue of nearly 20,000 infrared sources, mainly cool stars. Further refinements of infrared detection techniques and the use of ground-based telescopes at high, dry mountain sites provided a wealth of new information about the cosmos. A new window in the electromagnetic spectrum was open to the observation of astronomical sources. The first discovery of infrared astronomy directly relevant to our story came from the work of John E. Gaustad and his colleagues confirming that the infrared spectra of highly reddened stars showed no evidence of water ice (*Astrophys. J.* **158**, 151, 1969). Then followed a spate of other discoveries all showing a spectral feature of dust over the 8–12 μm mid-infrared waveband. The feature was observed in emission in a wide variety of astronomical objects including oxygen rich cool stars and in the Trapezium region of the Orion Nebula. (See for example, Fig. 2.) The results

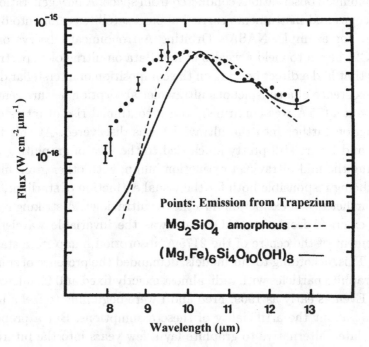

Fig. 2   The infrared emission by heated dust in the Trapezium nebula observed in 1969 (points). Curves show the behaviour of amorphous and hydrated silicates. The mismatch evident over the 8–9 μm micrometre waveband led us to consider alternative organic compositions of the dust.

were published in the form of several papers in the same issue of the *Astrophysical Journal Letters* in 1969. The authors E.P. Ney, F.C. Gillett, J.E. Gaustad, W.A. Stein, N.J. Woolf, R.F. Knacke, D.A. Allen and R.C. Gilman worked at various telescopes operated by the Universities of California and Minnesota. One of the papers in this volume interpreted the new results as evidence that the dust was made of a mixture of silicates — combinations of magnesium, silicon and oxygen as they occur in the rocks of the Earth. Their conclusion, however, was premature.

The papers in *Astrophysical Journal Letters* were soon followed up by a report by Nick Woolf and his colleagues describing the discovery of the same 10 μm feature now seen in absorption against an extended infrared source located at the centre of the galaxy. This was of course a clear indication that the feature in question was not confined to localised sources of dust around stars, but was a property of interstellar dust along a 10 kiloparsec path length to the centre of the galaxy.

I believe that Fred was given a few week's advance notice of the appearance of these papers by his friend Ed Ney, who was at the time the chairman of the astronomy department of the University of Minnesota. He asked me whether I thought silicate grains could fit the astronomical data that we had been modelling now for some years. What struck me immediately on studying the *Astrophysical Journal Letters* papers was how poorly the newly observed astronomical feature at 10 μm actually fitted the behaviour of any known silicate or mixtures of silicates. Since mineral silicates also have absorptions spanning the 8–13 μm waveband, there was of course a crude match to be seen. The "best" silicate fits to the Trapezium data are shown in Fig. 2 as curves for amorphous and hydrated silicates. But matches of this quality were not by any means restricted to silicates. Other chemical systems, including some that involved carbon, could be candidates. Whilst one could not dispute that some quantity of silicate dust might exist in space, how would this compare with contributions from the more cosmically abundant element carbon?

Our search for a carbon-based interpretation of the 10 μm interstellar feature began as early as 1969. An examination of spectroscopic literature showed that soot containing hydrogen admixtures had spectra that might be satisfactory from this point of view. Fred and I next calculated the behaviour of two component mixtures comprised of silicates and impure graphite grains, and showed that they could, at least approximately, fit the entire range of astronomical observations that were then available. A description of this work was published in the form of an article in *Nature* (*Nature* **223**, 459–462, 1970).

We were, however, aware of some glaring shortcomings. The fit of our synthetic model spectra to the new mid-infrared data in the 10 μm waveband, although better than for silicates, still left much to be desired. With the quality of the observational data that was now available, particularly in the extinction curve of starlight over the visual and ultraviolet bands, a more perfect agreement with models was required. There was also the question we referred to earlier, of justifying the artificial requirement of perfectly spherical graphite particles all of one size. And finally, how could we account for the almost invariant shape of the visual extinction curve? With a view to resolving at least some of these questions we began to consider three component, rather than two component grain models — particles comprised of a mixture of graphite, silicate and iron grains. Better results emerged, still not perfect. (Wickramasinghe and Nandy, *Mon. not. R. Astron. Soc.* **153**, 205–227, 1971). Increasing the number of constituents in our model with relative proportions that had to be fixed arbitrarily was also a source of concern.

For the next couple of years Fred's direct involvement with this programme of work had to be curtailed by the many strenuous commitments he had undertaken. Whilst the old cosmological wrangles still continued, Fred also became involved in matters relating to national astronomical policy. The volume of committee work that this involved robbed him of time for research and diminished his enormous creativity. But most soul destroying of all for him were the negotiations that were under way with Cambridge Universtiy and the Government for renewing support for his Institute of Theoretical Astronomy. Fred had quite definite views as to how it should continue into the next quinquenium, others amongst his adversaries

held different views. Despite the enormous success of the Institute
in a brief span of four years we are told that its very continuity was
being threatened as early as the summer of 1971. Fred relates this
part of the story in great detail in his autobiography "Home is where
the wind blows". I shall not reiterate the details, but merely say that
things turned out to be so intolerable for him that he submitted his
resignation to the University in January 1972.

The news came as a shock when it finally reached us in the early
Spring. Cambridge at this time was looking at its seasonal best.
The colleges were shedding the austerity and grandeur of their winter identity and the backs were now softening to a new ministry of
snow-drops, crocuses and daffodils. Stone courtyards, grey too long,
glimmered again in sunlight, revealing their glorious red and purple
depths. It seemed a cruel fact that ill winds could blow through so
exquisite a setting. For the next few weeks I felt a little numbed at
the thought that great changes were in the air. Dining with my wife
at a Ladies Night feast at Jesus College in April 1972, I wondered
if this would be my last such experience! Such feasts at Cambridge
Colleges (there were several each year) were unique social occasions,
lavish to a certain extent and mostly supported by endowments that
could not be used for another purpose. Feasts aside, I had bene-
fited from the many privileges that my decade-long Fellowship had
afforded. But most of all I had enjoyed the privilege of working so
closely with Fred Hoyle.

Chapter 9

The Cardiff Era

I spent most of the summer of 1972 in Sri Lanka with my wife and family. The era of mass communication had not yet dawned — faxes, emails and the internet were nearly two decades away and international dialling was expensive. So this chit of an isle in the Indian Ocean seemed blissfully but strangely remote.

From an early age the sea held an irresistible attraction for me. This is true also for very many people. The theory that life sprang from oceans, accounting for humanity's fascination with the sea, I was later to refute. Fred would always favour the exigencies of rugged landscapes as a means to reflect and meditate, but I would always choose the beaches and the ocean where communion is instantaneous. Wave after wave rolled forward from the distant ocean, interrupted only by the occasional rumbling of a train passing behind me. And vexations would momentarily recede, drowned by the great swell of the sea. Every evening the sun would set around 6.30 pm. The period of sunset that follows is brief because we are so close to the equator. The sunset colours depend very much on the disposition of the clouds. Sometimes the sunsets are spectacular, at other times a disappointment, mirroring the vicissitudes of life itself.

My thoughts this summer were co-mingled with nostalgia for my days at Cambridge that I feared were drawing to a close. It seemed also that my collaboration with Fred may have come to its logical end and I kept wondering what shape my career might take. Priya was pregnant with our second child, so thoughts about the future were even more resonant.

I returned to Cambridge that October and the reality of the situation dawned on me even more powerfully. As I walked into the Institute I felt a sense of sadness sweep over me, like returning to a house of bereavement — its life and soul had departed leaving what seemed an empty shell. Fred had left the Institute in mid August, never to return, and shortly afterwards sold his family home in Clarkson Road. Fred and Barbara had entered the next phase of their lives and were now in the process of resettling at Cockley Moor, near Pernrith in Cumbria in their beloved Lake District.

My own appointment at the Institute was assured only until the end of the academic year. What was to happen after that would be decided upon by the next Director. Without Fred the Institute lost its attraction for me and I could not see myself carrying on there, even if I were given the chance. The option of looking for openings in a Sri Lankan University was one I had considered but did not entertain seriously for too long. I was still fired with enthusiasm to carry forward the many promising lines of research I had begun, and the scientific culture of Sri Lanka could not permit the funding of such ventures. The country had more pressing economic problems to deal with.

I then began to look for university jobs in the UK. One of the first that came to my notice was the Chair of Applied Mathematics and Mathematical Physics at University College, Cardiff. Fred had not yet left the UK astronomical scene, so I was able to consult him on this matter. He strongly advised me to apply. I did so and my application was successful. At the interview for selection the Principal of University College Cardiff, Dr. Bill Bevan, made it clear to me that he was very keen to start Astronomy in Cardiff, and furthermore that he would like me to involve Fred Hoyle in this venture. I could not have expected a more attractive offer, so I accepted the position and agreed to assume duties in the summer of 1973.

My remaining few months in Cambridge were occupied with completing my book *Light Scattering Particles for Small Particles with Applications in Astronomy* that I had started to write earlier in the year. A major part of the book was a set of tables and graphs intended to save astronomers the tedious job of computing optical cross-sections each time they were needed in a problem. The book appeared in 1973 (*Light Scattering Particles for Small Particles with Applications in Astronomy*, John Wiley, 1973). For several years this book was much used by the growing number of workers on interstellar dust who did not have immediate access to the relevant programmes and high speed electronic computers. This situation changed dramatically in the 1980's with the advent of affordable personal computers, and later the internet. This type of information is now readily accessible on the internet, although I believe my book is still being used in some circles and occasionally even referenced. The one person for whom my "big red book" was immensely useful was Fred Hoyle. Fred could never be persuaded to use a PC let alone even consider using the internet. A hand held programmable HP calculator and a fax machine were as far as he could be pushed to acquire in the early 1980's.

Early in the summer of 1973 we sold our house in Barton Cambridgeshire and moved to Cardiff with our two young children. Our new home in Lisvane, a suburb of Cardiff, overlooks gently undulating mountains — a refreshing change from the flatness of the Fens around Cambridge. At the University I felt at first a little overwhelmed by the responsibilities entrusted to me. The task was by no means easy. There was jealousy and hostility of the existing staff of the department to contend with, particularly so as I had set out to steer the department in a direction that appeared quite alien to them. However, with the unstinting support of Principal Bill Bevan and senior members of Senate I was able to institute great innovations. Within a year of my appointment I had changed the name of the Department from "The Department of Applied Mathematics and Mathematical Physics" to "The Department of Applied Mathematics and Astronomy" and we were soon on the way to appointing 4 new lecturers in Astronomy. I had a dilemma of deciding upon the research fields in which to choose the new lecturers. Had I opted to have all appointments in the area of interstellar dust we would now probably have had the most powerful research group in the world in my own specialist field. Following the example of my mentor Fred Hoyle and the experience of the Institute in Cambridge I opted instead to develop as diverse a group as I possibly could. The first appointee was in the area of plasma physics, the second in the chemical evolution of galaxies, the third in star formation theories and the fourth in relativistic astrophysics. A year later I obtained support for a Chair of Observational Astronomy, so a highly diverse and balanced group was begun. The choice as I saw it in 1973 was like the difference between sowing a packet of mixed flower seeds and one of a single species of flowering plant. The mixed seeds if they took would lead to a glorious splash of colour. This is what has now happened in Cardiff, leading to the evolution of an astronomical research centre that is one of the best in the UK. Such heterogeneity is much rarer in the present day as subjects have become so overly specialised that people from different fields barely speak the same language as each other, and often cannot even recognise that they are trying to address the same problems.

By the late Spring of 1974 I had set all the major changes in train and was headed again to North America for a short respite — this time to Canada and to the University of Western Ontario in London, Ontario. Priya and I instantly took to Canada which we related to much better than its neighbour America. The country was more akin to Europe in terms of its culture and value systems, thus we found it more recognisable. It was during my three-month stint in Canada that I made a breakthrough that was to become a defining moment in this story. I mentioned earlier that we had not been satisfied with the quality of the fits that had been obtained with silicate models for the 8–13 μm astronomical spectra. So I was still searching for a better solution.

We had already found that hydrogen impurities in soot could offer a partially improved solution, but the absorption strengths in such bands were exceedingly weak. In any case it was clear that if an alternative to the then-fashionable silicate explanation was to be credible, the fit to the data had better be close to perfect. As these thoughts were running through my head, I was beginning to feel the 8–13 μm band mismatch of the silicate model must somehow conceal a much bigger story. The carbon with hydrogen model described in our 1969 paper was in a sense skirting around the possibility of an organic grain model. What if the carbonaceous component of the dust was not simply graphite but made of organic materials, organic polymers in fact? The ice grain model of van de Hulst was argued as a plausible composition because in principle at least oxygen could combine with hydrogen to form the stable molecule of water. In the same way carbon atoms in interstellar space might be combined with hydrogen and oxygen to form an extraordinarily vast variety of organic chemicals. In terms of the basic chemical elements at least there would be more than enough mass to explain the properties of interstellar dust.

At this time the molecule formaldehyde H_2CO had been discovered to exist ubiquitously in interstellar clouds. It was present in dense molecular clouds as well as in the less dense interstellar medium. What if such molecules started to condense and polymerise on the surfaces of pre-existing graphite or silicate grains expelled from stars? Looking though the books in the library of the University of Western Ontario dealing with properties of formaldehyde, I soon discovered that it could readily polymerise on the surfaces of silicates. It also turned out by a stroke of luck that the Chemistry Department here had one of the leading experts on formaldehyde polymers and I had the benefit of many discussions with him. A simple calculation showed that under interstellar cloud conditions substantial mantles of formaldehyde polymers would indeed grow. Next I began to look at the optical and infrared properties of many types of formaldehyde polymers. It turned out that such polymers were dielectric (that is to say, non-absorbing) in the visual waveband as the observations demanded. And most strikingly polyformaldehyde (polyoxymethylene) had absorption bands over the 8–12 and 16–22 μm wavebands, with the former absorption fitting the astronomical data better than any known silicate, as shown in Fig. 3. This was a breakthrough moment and within a couple of weeks my paper entitled "Formaldehyde Polymers in Interstellar Grains" was submitted to *Nature*. It was published (*Nature* **252**, 462, 1974) and made a splash in the science news columns of the broadsheets. This was the first ever suggestion of the widespread occurrence of organic polymers in the galaxy, and it was also the first paper to come from the newly reconstituted Department of Applied Mathematics and Astronomy in Cardiff.

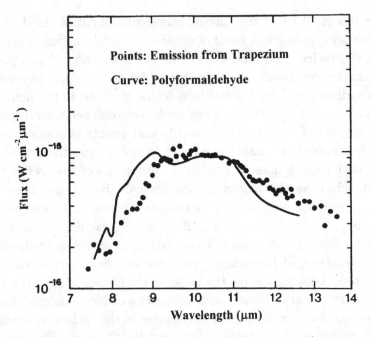

Fig. 3   The infrared emission from the Trapezium nebula (points) compared with the behaviour of polyformaldehyde dust (curve) heated to 300 K. This was the first argument used in favour of organic polymeric dust in 1974.

Within days of the publication of my paper, Professor V. Vanysek of Charles University, Prague came to visit me with a proposition to be considered. Polymerised formaldehyde and also other organic polymers, could form a major component of comet dust as well. Comets were thought of at this time to be dirty snowballs, mostly comprised of inorganic ices with siliceous and metallic dust occurring as minor impurities. When a comet approaches the inner parts of the solar system, the molecular species in the developing coma were thought to be break-up products of ices which were regarded as "parent molecules." With polymerised formaldehyde being a component of the comets, coma radicals such as OH, CN, C_2 could be interpreted as the break-up products of such a polymer. When we began to explore this idea together it appeared to us to seem more and more plausible. Since formaldehyde polymers remain stable up to temperatures 500 K, changes in the 10 μm spectrum of the new

Comet Kohoutek (1973f) as it came within 0.5 AU of the sun were shown to be consistent with this model. The idea of an organic comet was thus born, and its justification was described in a joint paper published in *Astrophysics and Space Science* (V. Vanysek and N.C. Wickramasinghe, *Astrophys. Space Sci.* **33**, L19–28, 1975). This publication established our priority for the idea that both interstellar and cometary dust could be comprised of organic polymers.

Up to this point in my story there was no major discord with mainstream astronomical ideas, and indeed there were even some grudging plaudits being accorded to us for making such an innovative suggestion. I soon acquired a research student, Alan Cooke, who started doing more detailed modelling of formaldehyde grains, and a grant from SERC (the Science and Engineering Council of the UK) to do experimental work on the optical properties of organic polymers was also secured. From 1975 to 1977, I was awarded more grants from SERC that enabled several visitors to be invited to Cardiff for collaborative work.

Chapter 10

The Search for Cosmic Life

Two years after I began my tenure at University College Cardiff, I succeeded in getting Fred appointed as an Honorary Professor. This meant that he would be a nominal member of my Department and be entitled to use the university as endorsement for his publications. In turn, Cardiff gained much kudos from this link with Fred Hoyle. During the period of 1975–77, Fred spent large chunks of his time in the United States, so that his appearances in Cardiff were rare. Even so, I usually managed to track him down and keep him abreast of our research developments on organic polymers.

Most of our efforts in Cardiff were now directed at securing optimal fits to the 8–13 μm emission feature in one particular astronomical source — the Trapezium region of the Orion Nebula. (See Figs. 2 and 3.) We began to regard this source as the touchstone of acceptable grain models. (Modelling of this source was particularly simple because it was optically thin in the infrared, with grains all emitting at a temperature within the range 150 to 300 K.) A most glaring inconsistency with silicate spectra showed up as a deficit of silicate emissivity over the wavelength interval from 8 to 9 μm. The difficulty appeared to be endemic to all types of silicates, amorphous and hydrated. Moon rock was also considered for comparison, as were silicates that had been irradiated in the laboratory with a view to simulating space conditions. But nothing seemed to work!

A trick was invented by astronomers who thought they had to stick with a silicate hypothesis. The normal logic of identification through spectroscopy was inverted. The astronomical spectrum of

77

the Trapezium Nebula was used to infer the opacity properties of the
emitting material. A hypothetical material with such opacity properties but which has no counterpart in the real world was named
an "astronomical silicate". We considered this practice unsatisfac-
tory in the extreme and openly deplored it as being a "cheat". The
fact remains that no known silicate can account for the Trapezium
data as well as the organic polymeric models we discussed in the last
chapter. There was what seemed to us to be unequivocal evidence
of organic polymers existing on a vast scale throughout the galaxy.
Models involving co-polymers — mixed chains of formaldehyde and
other molecules as well as polymer mixtures resembling tars — were
leading inexorably in one direction — life. What if the interstellar
grains that I had begun investigating in 1960 were indeed connected
with biology, with life itself? This question, with all its profound
implications, represented such an assault on conventional thinking
that I felt a compulsion to involve Fred at this stage.

I began a correspondence with Fred early in 1976 by first suggesting that the polymeric grains in molecular clouds could represent the
beginnings of a process that may lead to life, thus permitting life to
originate and evolve on a much bigger scale than had hitherto been
contemplated. Fred's first reactions were far more cautious than I had
anticipated. This point is worth stressing because there is a general
perception that he embarked on our joint projects rashly and uncriti-
cally. Nothing can be further from the truth. He was exceedingly critical of every radical proposition that was put to him at each stage in our
collaboration. He played the role of devil's advocate until he was convinced that there were overwhelming arguments to support the radical
proposition. And this is exactly how a true scientist should proceed.

After many exchanges of letters and papers, Fred and I agreed
in August 1976 that organic polymers in grains could undergo a
Darwinian-style prebiological or prebiotic evolution in grain-on-grain
collisions during the collapse of a molecular cloud. Organic tarry
grains tend to be sticky and grain clumps would form by particles colliding and sticking together. Such grain clumps would also
trap other organic molecules from the gas, and chemical transfor-
mations, sometimes assisted by ultraviolet light, would take place in

the condensed state. In our first paper referring obliquely to biology entitled "Primitive grain clumps and organic compounds in carbonaceous chondrites" (*Nature* **264**, 45–46, 1976) we wrote:

> "*The formation of simple amino acids (e.g. glycine) is expected to take place in dense molecular clouds which may well be the cradle of life.*"

Even such a tentative proposition was regarded as outrageous heresy in 1976, although now in 2004 it is regarded as obvious.

Towards the latter part of 1976 there were other events that were to have a bearing on our story at a much later stage in its development. Even at the risk diverting from the main thread of my argument I shall report one such event here if only to keep my record in an approximate chronological order.

Space exploration was gathering momentum, as well as the search for life in the solar system. There were two unmanned missions to Mars: Viking 1 arrived at the red planet on 20 July 1976 and Viking 2 on 3 September 1976. Each mission involved an "orbiter" that was set in motion around the planet and a "lander" that actually set down at a chosen spot. The two Viking landers arrived safely at their chosen destinations, equipped with apparatus to make *in situ* tests for the presence of microbial life. These experiments were of vital importance and in many ways much more explicit in detection of life than any subsequently contemplated.

In one experiment (designated LR) a nutrient broth (with an isotope label on the carbon) of the sort that is normally used to culture a wide range of terrestrial microorganisms was contained in a sterilised flask, and the Martian soil was robotically added to it. It was found that the nutrient was taken up by the soil and gases (CO_2) were expelled from the flask as would be expected if bacteria were present. In another experiment the soil sample was heated to 75°C for 3 hours before it was added to the nutrient. This led to a diminution of gas release by 90%, but significantly the reaction was not completely stopped. Since some bacterial and fungal spores could survive temperatures of 75°C, the result of the second experiment was also consistent with a biological explanation, especially as the activity recovered gradually to its former higher value as time went on. The bacterial explanation gained further support from a third result, obtained by heating the soil sufficiently to kill microbial life entirely, when all activity was found to stop. However, yet another experiment in the Viking package proved initially more difficult to reconcile with biology. This experiment, designated GCMS, sought to analyse the organic content of Martian soil using a mass spectrometer. Here the results were disappointingly negative for organic matter, indicating that if such matter existed it was present only in very small quantities.

The fact that the LR experiment was decisively positive and the GCMS experiment was negative posed a difficulty for NASA. The outcome was indefinite, and this is the way it should have been presented to the public. Yet NASA elected in 1976/77 to announce that the Viking experiments did not support the presence of life, and their statement that Mars was an intractably lifeless planet was given a great deal of publicity in 1976. It was their view that some other non-biological explanation had to be sought and would eventually be found. This has not happened to date and it has to be conceded that the balance of evidence is still positive for Viking landers to have detected life. But even retrospectively, there still remains an enormous resistance to admit this, and modern pronouncements by NASA relate mostly to the presence of past life in an epoch when rivers flowed over the surface of the planet. The facts have been clouded over, perhaps in an attempt to hide the fact that they plumped for an incorrect theory. And when big government science and the media make errors together, neither is anxious to be seen correcting itself, a sufficient reason why nothing much that is good can come of public funded science done in the glare of publicity. Science is a quiet, reflective activity, which cannot flourish in modern egalitarian or totalitarian societies.

The principal investigator of the 1976 biology experiment on Viking, Gill Levin, is my friend, and has revealed many things that are not generally known to the public. For instance his studies of a sequence of pictures taken by the Viking cameras over the duration of a Martian year, showed subtle shades of green appearing on the tops of rocks in the spring. These receded in the winter, suggesting the growth of lichen-type microbial life. Levin's views about these findings, however, did not endear him to the administration of NASA, and he parted company from them shortly afterwards to pursue his investigations independently. And after several decades of experimentation it has turned out that no non-biological model is feasible for explaining the positive results of the LR experiment, and moreover the lack of free organic matter in quantity as revealed by the GCMS experiment can easily be explained on the basis of a slow turn-over rate of microorganisms to be expected under the relatively inhospitable conditions that prevail on Mars.

The Mars probe Odyssey was launched in April 2001 to orbit the red planet and map its surface for hydrogen, water and minerals. Named *Odyssey* after Arthur C. Clarke's blockbuster novel, the probe obtained pictures that showed clear evidence of heavy frost or snow in many locations including the Viking landing sites. Snow or frost deposits were found to be seasonal, pointing to some kind of water cycle. But still NASA was claiming that contemporary life was highly improbable, despite the fact that on many sites on Earth where life has been discovered — in antarctic ice and at depths of 8 km below the Earth's surface — there are unquestionable parallels with Mars. Even as recently as 2004, the spacecraft "Mars Express" has obtained traces of methane and oxygen in the atmosphere that together would normally be interpreted as indicating biological activity. NASA seems loath to publicise such findings, perhaps in the hope that if they present the case very slowly, they would then be able to claim exclusive priority for the discovery of extraterrestrial life. Arthur C. Clarke summarised the four stages of the way new ideas are accepted into mainstream, institutional science:

- these ideas are crazy, do not waste my time with them;
- these ideas are possible, but are of no importance;
- these ideas — we said they were true all along;
- these ideas — we thought of them first.

I shall now return to the story of the composition of the interstellar dust. I had been plagued for a while by the thought that graphite grains, for which idea I had by now acquired a degree of fame, could not offer a rational explanation of the 2175 Å interstellar absorption band. The requirement of spherical graphite grains, all of one radius, 0.02 μm, was difficult if not impossible to defend. The central wavelength of the absorption due to graphite shifted away from the astronomically observed wavelength for particles in the shape of flake or whiskers, or if the radii of spheres departed significantly from 0.02 μm. Whilst continuing to fine-tune agreements with the 8–30 μm astronomical spectra in the infrared waveband with various types of organic polymers, it occurred to me to look critically at their ultraviolet spectra as well. In the summer of 1976 I discovered that a significant class of organic materials that possess C=C double bonds in their structures have ultraviolet spectra that peak near the required wavelength. Moreover from the available laboratory measurements in the ultraviolet we could calculate that only 10% of the available carbon in space was needed in the form of this material to give a 2175 Å band of the observed strength. This was my next subject of correspondence with Fred. Within a short while he agreed to our publication in *Astrophysics and Space Science* (**47**, L9–L13, 1977), the authors being myself, Fred Hoyle and Kashi Nandy of the Royal Observatory Edinburgh who was involved in observation of the interstellar extinction curve. This publication carried the first exposition of the idea that the ultraviolet extinction band was due to complex organic molecules.

The year 1977 was a vintage year for our collaboration, with no less than 6 papers being published in *Nature*. We were moving steadily in the direction of astrobiology, possibly 20 years ahead of our competitors. It should also be noted that we were remarkably successful in airing most of our work in the so-called "high impact" journals. The campaign of outright censorship had not yet begun.

In the world at large outside science there were momentous events afoot: in Britain the Queen celebrated her Golden Jubilee, Space sensationalism glutted Hollywood with movies like "Star Wars" and "Close Encounters of the Third Kind", almost as though the populace at large was getting ready for the arrival of ET in some form! In my native country, Sri Lanka, Junius Jayawardene becomes Prime Minister.

In January 1977 Fred made his first long visit to Cardiff in his capacity as Honorary Professor. During this week-long stay, he resided with us in Lisvane, as he would do on numerous subsequent occasions over the next decade and a half. It was always a pleasure to entertain Fred, and as the years went by, we became increasingly relaxed in his company. Priya is a charming and peerless hostess, and her efforts, particularly her cooking, were greatly appreciated by Fred. In fact she later came to joke with him about her contributions to man and science. Our friends, associates, and many others, were inevitably keen to meet Fred when he was in Cardiff, so places at our table were in high demand! We would invite small groups of guests to dinner, and Fred was charming company on such occasions, and his anecdotes were varied and surprising. I remember that once he nonchalantly related how, when he had finished the script for *A for Andromeda*, he toured the reparatory theatre in search of a "boyish young girl" to play the lead role. He was arrested by a performance of an unknown actress, Julie Christie, who agreed to play the part. By the second series, however, she had acquired such a degree of fame, that she was too expensive to re-hire.

As far as our collaboration was concerned Fred's visit in January 1977 was mostly taken up with the task of poring over vast tomes of atlases of laboratory infrared spectra of organic polymers. Our comparison to be made was with the spectrum of the so-called BN object in the Orion Nebula. (See Fig. 4.) One of the defining features of organic polymeric dust would be a $3.4\,\mu m$ feature of C–H bonds. In astronomical sources studied so far this feature shows up only as very weak shoulder on the wings of a much more prominent $3\,\mu m$ absorption band. The $3\,\mu m$ band was interpreted at the time as being due to H_2O ice condensed on grains within the dense molecular cloud around the BN object. Looking at the best available astronomical spectra Fred and I were the first to spot the CH absorption effect quite clearly, albeit as a shoulder on the longwave wing of an "ice" band. The task we set ourselves now was to find an organic polymer with absorptions at 2.9–3.1, 3.4 and 8–$13\,\mu m$ that was able to produce a BN-like spectrum.

This was not easy for the reason that for most of the organics we looked at, the CH stretching band at $3.4\,\mu m$ tended to be much

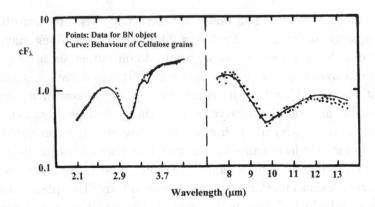

Fig. 4 The first argument for an interstellar biopolymer made in 1977. The infrared emission from the BN object in the Orion Nebula (points) matched to a model involving cellulose (curve).

stronger than the OH stretching band at about $3\,\mu$m. We were, however, able to argue that there were both water-ice and organic polymers associated with the BN source. Because the mass absorption coefficient of water ice at $3.1\,\mu$m is a thousand times that of the $3.4\,\mu$m band for most organic polymers, a mass ratio of ice to organics of 1:1000 would produce the hint of a longwave shoulder in the $3.1\,\mu$m absorption band, exactly as it is observed in astronomy. In other words only a trace of water needs to be present in the dusty material around the BN object that had to be overwhelmingly organic. All these matters were sorted out during Fred's visit to Cardiff in January 1977.

We also began to look now at new ultraviolet spectra obtained for an extract of organic molecules from the Murchison meteorite. The data was supplied to us by A. Sakata in Tokyo, and we could immediately see here that the spectrum possessed a mid ultraviolet absorption feature near 2175 Å, very similar to the observed interstellar extinction feature to which we have referred earlier. We submitted a short letter to *Nature* making this point which confirmed our earlier contention of an organic carrier for the 2175 Å interstellar band. (A. Sakata *et al.*, *Nature* **266**, 241, 1977).

Our collaboration was rapidly gathering momentum. We entered a phase involving a brisk exchange of telephone calls, letters and graphs between Cockley Moor and Lisvane. We had eventually stumbled upon an organic material that captured our interest. It was an infrared spectrum of cellulose. The laboratory spectrum of cotton cellulose over the 2.5–30 μm waveband had shown even at a cursory glance most of the features required in order to explain astronomical spectra such as BN (see Fig. 4) and also the Trapezium nebula. The new agreements we obtained were somewhat better than those obtained earlier for polyformaldehyde as seen for instance in Fig. 3. Cellulose is of course the main component of the cell walls of plants and is by far the most abundant terrestrially occurring biopolymer. It is however logically at least the simplest biopolymer with an empirical formula $(H_2CO)_n$ just the same as for polyformaldehyde. It is a member of the most stable of a set of polymers known as the polysaccharides, which involves chains of various types of sugars, with a pyrolysis (heat destruction) temperature of 800 K. The advantage of cellulose is that it could exist in regions of relatively high temperature such as HII regions (ionised hydrogen), of which the Trapezium nebula in Orion is an example.

My own instinct at this stage was to consider interstellar polysaccharides as being derived from life and being indicative of fully-fledged microbial life in the Galaxy. Fred, however, was still inclined to tread more carefully. He conceded the existence of polysaccharides and similar molecules in interstellar space, but still sought non-biological or abiotic processes for their formation.

Both Fred and I spent a lot of time modelling a wide variety of astronomical observations over the 2–4 μm and 8–14 μm wavebands using the opacity measurements for cellulose. These efforts led to the publication of a series of papers (e.g. *Nature* **268**, 610–612, 1977; *Mon. not. R. Astron. Soc.* **181**, 51–55P, 1977). A particularly impressive fit with cellulose was calculated for the source OH26.5 + 0.6 where data was available over the very long wavelength interval from 2 to 40 μm (Fig. 5). Fred regarded this fit as the most decisive evidence for the validity of our point of view as it was in 1977.

The modelling of all these sources required a straightforward procedure known as a radiation transfer calculation which was a trivial job on the Cardiff University computer. But Fred insisted on checking

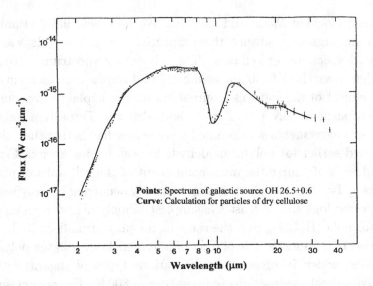

Fig. 5   The spectrum of the galactic infrared source OH26.5 + 0.6 over the very long waveband (2.5–40 μm) compared with a cellulose model (curve).

everything himself and he performed all his calculations on a simple hand-held programmable HP calculator that he carried everywhere. He maintained that being close to the logic of a calculation gave him a better insight into what was going on when it came to assessing the significance of the solutions that were obtained. Nowadays, school children use calculators and computers even for the simplest arithmetical operations, and this practice has undoubtedly led to a decline in the standards of numeracy, with multiplication tables being a thing of the past.

In parallel with our work on cellulose to model infrared spectra we also took another crack at the problem of understanding the interstellar ultraviolet absorption. This was an extension of the work on the Murchison meteorite extract that we referred to earlier. We argued that a class of double-ringed (bicyclic) aromatic compounds with the empirical formula $C_8H_6N_2$ (Quinozoline, Quinoxaline) provide an alternative explanation of the interstellar 2175 Å absorption band (*Nature* **270**, 323–324, 1977). A comparison of the astronomical data and the average absorption properties of these compounds is shown in Fig. 6. Our paper contains the first suggestion in the literature

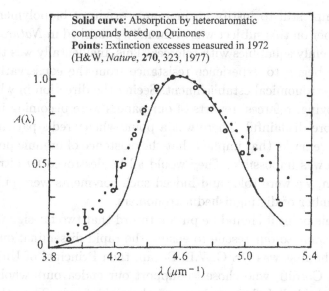

Fig. 6 The first argument for polyaromatic hydrocarbons in space in 1977. The average spectrum of compounds based on quinones (curve) compared with the 2200 Å interstellar absorption feature as it was known in the 1970's (points).

of interstellar aromatic compounds and accords unequivocal priority to Fred and myself for the idea of the existence of interstellar polyaromatic hydrocarbons — a presence that is now taken for granted. This work was published in an accessible and credible journal, and so we found a lack of referencing to it inexcusable by any standards of moral propriety. I shall refer to similar matters in a later chapter.

To my mind the identification of polyaromatic hydrocarbons and cellulose-like polymers in interstellar space were tantamount to biology. Still preferring abiotic explanations Fred discussed a model involving mass flows from high mass O type stars, with mass loss rates amounting to about a hundredth of a solar mass per year. The outflowing matter then absorbs and re-emits the radiation from the star until an effective photosphere with a temperature (much cooler than the star) in the range 5000–10,000 K, forms at a certain distance. Beyond this point as the gas cools, one could argue that molecules form, including ring molecules and polysaccharide chains. This model was ingeniously conceived by Fred and worked out in great detail in an attempt still to justify an inorganic origin of biopolymers. Our joint paper on this subject was eventually published in *Nature* after a few unseemly squabbles with referees. This, incidentally was the first time we began to experience resistance from the conservative and hostile astronomical establishment. Seeing the direction in which we were moving, referees' reports of our papers were becoming increasingly more disdainful. There was a point when Fred's patience was tried by remarks that implied that the existence of organic polymers in space was impossible. They would all be destroyed by ultraviolet radiation, we were told, and indeed such comments were published by a number of distinguished astronomers.

At this point it should be put on record that two far-sighted individuals came to our rescue to ensure the rapid dissemination of our work. The first was Dr. C.W.L. Bevan, then Principal of University College, Cardiff, who chose to support our endeavours wholeheartedly in the belief that we were on the right track. The other was Zdenek Kopal who was the founding editor and Editor-in-Chief of the journal *Astrophysics and Space Science*, and at the time Professor of Astronomy at Manchester. Kopal offered us the opportunity of publishing our ideas in his journal, whilst Bevan agreed to the funding by University College Cardiff of a "Blue Preprint Series". The now extinct University College Cardiff Press was essentially placed at our disposal for publishing preprints as well as monographs on subjects of our choice. In this way our channels of communication with the scientific community were not interrupted as the result of the unexpected turn that our researches were to eventually take.

It was at about this time that we began to feel the need for a dedicated laboratory facility to obtain spectroscopic data as and when we required them. Whilst still exploring the role of interstellar polysaccharides we realised somewhat belatedly that the Biochemistry Department of University College Cardiff headed by Ken Dodgson had on its lecturing staff a leading expert on polysaccharides, Tony Olavesen. We wasted no time to seek his help to measure for us the spectra of a large number of different polysaccharides under conditions we considered were appropriate for making comparisons with astronomy. This work soon gave us confirmation of our earlier conclusions that were based only on the published spectra of cotton cellulose, and led to another paper that was published by *Nature* (F. Hoyle, A.H. Olavesen and N.C. Wickramasinghe, *Nature* **271**, 229–231, 1978).

By now it was becoming clear that the scientific establishment had had enough of our irreverance for authority. I felt we were walking into a kind of scientific exile. Even though we were still sticking firmly to prebiotic molecules in space a surge of resentment was beginning to surface. Fred's small hand-held programmable HP was being stretched beyond its limit for the ambitious calculations he was intending to do. Fred and I therefore decided to apply to the Science Research Council (SRC) for the purchase of a small computing facility to be housed at Cockley Moor for this purpose. The following reply came from S.T.G James of the SRC dated 4 November 1977:

> *"Dear Professor Hoyle,*
>
> *Your joint application with Professor Wickramasinghe for a grant a £10,540 in support of computational work on the identification of polysaccharides and related organic polymers in galactic infrared sources, was considered by Astronomy II Committee on 25 October.*
>
> *I regret to have to inform you that the Committee were not prepared to support the proposal outlined in your application. In arriving at this decision the Committee indicated that it was not at all clear that the problems mentioned in the application were ready for attack by computational methods . . . ".*

This response was an indication that our work on interstellar prebiotic polymers was being firmly rejected by the astronomical establishment in 1977. In retrospect the irony here is that these ideas are now firmly in the mainstream, with little credit being recorded in our favour. The gross inequity of the rejection of our grant application made us take an unusual step. We wrote to the Prime Minister James Callaghan, who was also an MP for Cardiff. A reply dated

7 February eventually came from the Secretary of the Department of Education and Science with the following pronouncement:

> "*It appears that this application has been properly assessed under the regular procedures of the SRC and that you have been given reasons for rejection...*".

All this did not deter us from proceeding along the course upon which we had embarked. At this time Fred was firmly wedded to the idea of stellar mass flows and interstellar chemistry producing organic molecules. These were the complex structural building blocks of life that were supposed to become incorporated into the comets of the early solar system, within which an origin of life in the fashion of the Haldane–Oparin model took place. I was, however, personally inclined to flirt even more confidently with the radical idea of interstellar microbial life, interstellar polymers being regarded as having a biological provenance.

Fred was in the US again when the next crucial step of our journey was taken. I received a letter from Dr. J. Brooks of the School of Chemistry at Bradford University containing a spectrum of a complex organic biopolymer known as sporopollenin, which forms a major component of pollen and many spore walls. The spectrum had many of the features that were required to explain galactic infrared sources except for one fact. A 3.4 micron feature due to CH stretching was too prominent compared to the galactic sources we had seen. If, however, a thin layer of ice was condensed on particles comprised of sporopollenin, this difficulty would be rectified, thus making ice-coated bacterial spores a tenable model for interstellar dust. I promptly wrote to Fred in the United States soliciting his views and hopefully his support on this proposition. I soon followed this up with a draft of a joint paper for *Nature*. In a letter dated May 5 1977 written from Cornell University he wrote thus:

> "*Is the association of sporopollenin specific enough to support the final paragraph (speculating on the possibility of spores)? One might grant an interesting relation of the IR absorption obtained for the galactic centre or for galactic sources with*

the IR curve for sporopollenin, but the association may not require anything as complex biologically as sporopollenin itself...".

A few days later he reiterates the same point. In a letter dated May 9 also from Cornell he writes:

"My feeling, however, is that we cannot invert the situation toward a conclusion favouring a particular complex biological structure like sporopollenin:

Polymers → IR absorption	*O.K.*
IR absorption → Interstellar Biology"	*Not O.K.*

A drastically toned down version of the first draft I sent to Fred (that included references to interstellar biology) eventually appeared in October in *Nature* (Wickramasinghe, Hoyle, J. Books and G. Shaw, *Nature* **269**, 674–676, 1977).

Later that summer Fred was back home in Cumbria. Still using his hand-held HP he was calculating spectra of a large number of galactic infrared sources with opacity data for polysaccharides. The results, in terms of the closeness of fits, were most impressive. We first issued our calculations as a preprint in our "Cardiff Blue Preprint Series", and later in two papers — a short version in *Nature* (*Nature* **268**, 610–612, 1977) and a fuller version in *Astrophysics and Space Science* (*Astrophys. Space Sci.* **53**, 489–505, 1978).

In July of this year, Phil Solomon, one of the pioneers in the detection of interstellar molecules using millimetre waves, visited me in Cardiff. He met with Fred both in Cardiff, and also in Mid-Wales for a workshop on Giant Molecular Clouds in the Galaxy that I had organised. The venue for the symposium was Gregynog Hall, a Conference Centre belonging to the University of Wales. With links dating back to the 12th century, Gregynog Hall is an imposing 19th century manor house set amidst 750 acres of gardens, woodland and farmland. It was bequeathed to the University in the 1960's by the Davies sisters, who had been avid art collectors, as well as generous patrons of the arts. Fred and Barbara, along with their two grandchildren, rented one of the many cottages in the grounds. Priya and I, and our two young children, did the same. The kids were all of a similar age and enjoyed having free rein to tear around the vast grounds together. The conference itself brought together a small group of scientists, and was hailed as a success. Fred and I had ample time between sessions to discuss our strategy for further development our ideas. Our joint presentation "Evidence of Interstellar Biochemicals", was published in *Giant Molecular Clouds in the Galaxy* (eds. P.M. Solomon and M.G. Edmunds, Pergamon Press, 1980).

In September 1977, we were abroad again, visiting the Astronomy Department of the University of Western Ontario, Canada. Fred was on his way to Caltech. Perhaps in need of a respite from the intensity of the life in space argument, Fred and I started poring over a volume entitled "Cretaceous-Tertiary Extinctions and Possible Terrestrial and Extraterrestrial Causes" published in the previous year by the Canadian Museum of Natural History in Ottawa. Some 65 million years ago the dinosaurs and indeed all animals with body weights above 25 kg suddenly became extinct. We argued that this could be due to the interaction of the Earth with a cloud of porous cometary dust derived from the extended coma of a comet. The Earth's stratosphere will be dusted over in a way that two thirds of the light and incident energy from the sun will be blocked for several years whilst still permitting infrared radiation to leak out. The result would be semi-darkness for a decade, leading to the withering of foliage in trees and causing a severe interruption of food chains. Herbivorous creatures including dinosaurs would soon become extinct, and so would carnivores that feed on the herbivores. With rivers still continuing to run and some lakes remaining unfrozen, fresh water organisms would survive — their food chains depending on decaying vegetable matter, would take longer to disrupt than marine organisms dependent on phytoplankton. The seeds and nuts of land plants would also survive and small animals, including small mammals, living on nuts and seeds would also survive the dark and desolate years. We humans owe our descent through this ecological crisis to the survival of these small mammals. All these ideas were published in the form of a paper in *Astrophysics and Space Science* (*Astrophys. Space Sci.* **53**, 523–526, 1978).

Our ideas on the Cretaceous-Tertiary extinctions were similar (though not identical) to those of Alvarez and Alvarez that were published approximately 2 years after ours, and which have come to be more or less generally accepted. A direct hit by a comet seems to have occurred 65 million years ago, and the resulting crater has been discovered in the seabed of the Yucatán Peninsula.

In addition to causing extinctions of species we also argued in our paper that cometary dusting over a more protracted period could trigger the onset of ice ages. This process was explored by us in greater detail in the late 1990's.

Chapter 11

Life from Comets and Pathogens from Space

Serendip is an ancient name for the island of Sri Lanka and was in use from the 4th Century AD. In the fairy-tale of Horace Walpole (1717–1797), *Three Princes of Serendip*, the heroes keep making delightful discoveries of things that they were not in quest of. This, according to the Oxford English Dictionary is the origin of the English word serendipity.

It is perhaps not a surprise that serendipitous events played a part in our collaboration at crucial stages of its development. This was certainly so for events that were to lead to a particular diversion that was to engage our attention almost obsessively for a full three decades. It is a curious tale that started with sniffles in the late summer of 1977, just prior to my trip to Canada. I had succumbed to an unseasonal bout of flu-like illness and this happened at a time when Fred and I were in a phase of brisk telephone exchanges over matters relating to the origin of life in comets. I was suddenly reminded of my mother's admonition in my childhood: "Don't go out in the rain or in evening mists or you'll get ill!" A similar belief is, of course, widely prevalent in the West. And in many cultures throughout the world comets have also been thought of as harbingers of pestilence and death. Could time-hallowed beliefs possibly have their basis in hard fact?

I telephoned Fred in Cockley Moor, on a depressingly grey afternoon in the late summer, with perhaps the most preposterous proposition I had ever made. Could the old wive's tales of diseases being connected with rain have possibly been right? Could viruses be present in comets, and could cometary viruses entering the Earth cause disease? Fred was caught unawares and I was greeted by a long silence at the other end, before he finally said he would think about it. I was indeed extremely surprised when he phoned back within hours of my call agreeing that this might be so. Fred was reminded of conversations he had had some years earlier with the Australian physicist E.G. (Taffy) Bowen. Bowen had discovered an amazing connection between freezing nuclei in rain clouds and the incidence of extraterrestrial particles.

What Taffy Bowen had showed was that there was a link between the frequency of freezing nuclei in tropospheric clouds and the occurrence of meteor showers. Meteor showers occur at regular times in the year whenever the Earth in its orbit crosses the trails of debris evaporated from short-period comets. If bacteria and viruses come in with cometary meteor showers they could, if they survive entry, act as freezing nuclei for rain. Raindrops laden with bacteria and viruses then become a distinct possibility. By analysing all the available data Bowen had reported in a paper in *Nature* (**177**, 1121, 1956) that as dust from meteor streams falls into the troposphere heavy falls of rain could be expected. This is found to happen some 30 days after the meteor dust first entered the very high atmosphere. It was discovered only much later that bacterial and fungal spores could indeed lower the freezing temperature of water and act as condensation nuclei for rain.

Because Fred was still hesitant in 1977 to accept the possibility of bacterial grains in interstellar space, his instant conversion to the idea of disease-causing bacteria and viruses coming from comets somewhat surprised me. We did not waste much time in issuing a preprint expounding our position, as we conceived it at the time, both on the origin of life in comets and on its consequences in regard to disease.

We began our exposition with the well-accepted premise that comets originated within our solar system as a by-product of the formation of the outer planets Uranus and Neptune. As I have previously discussed, a disc of gaseous material expelled from the equatorial regions of a fast rotating superluminous sun is the setting for the formation of all the planetary bodies. The inner planets condense as the disc cools to 1400 K thus removing metals and silicates from the expanding planetary disc. At the distances of the outer planets icy particles form which then coalesce into hundreds of billions of cometary bodies. This disc of comets remains immersed in the dense molecular cloud from the solar system condensed for hundreds of millions of years. Since this entire system undergoes random motions through the cloud, it would mop up through contact sweeping, tens of Earth masses of interstellar prebiotic molecules.

The aggregation of the outer planets took some 300 million years, after the inner planets had condensed into solid bodies. During this period cometary bodies would have been randomly deflected into elongated orbits that crossed the orbits of the inner planets. Several direct hits of the Earth by comets would have taken place early in its life and this led to the formation of the Earth's primitive atmosphere and oceans.

During close perihelion passages (closest approach to the Sun) of comets there would have been a tendency for selectively boiling off volatile materials in the nucleus. The surface temperature of the comet would oscillate between 300 K at perihelion and 100 K at aphelion (furthest distance to Sun) giving rise to periodic oscillations of melting and refreezing near the surface. We argued that such a process could lead over many millions of years to the elaboration of interstellar prebiotic molecules into more complex structures and eventually to life. A soft landing of a comet on the Earth about 4 billion years ago then could have started life. All this was to be the subject of our first book *Lifecloud: The Origin of Life in the Universe*, which was published by J.M. Dent and Sons in 1978.

But other attempts at starting life anew from prebiotic molecules, we argued, must continue on the surfaces of comets, particularly on "new" comets which have long periods of revolution around the sun. At irregular intervals the Earth must pick up debris from such comets in the form of micrometeorite showers; and at more regular and frequent intervals it picks up material from shorter period comets. The question naturally arises whether encounters with such material, containing fresh attempts at starting life, could sometimes have a deleterious effect on indigenous terrestrial biology. In answer to this question we began to argue that extraterrestrial encounters of this type may indeed lead to the injection of disease-causing bacteria and viruses on to the Earth.

By observing the pattern of appearance of diseases in the past we soon discovered that a strong *prima facie* case existed in support of our contention. The rather sudden appearance in the literature of references to particular diseases is significant in that it probably points to times of specific "invasions". Thus the first clear description of a disease resembling influenza is early in the 17th century AD. The common cold has no mention until about the 15th century AD. Descriptions of small pox and measles do not appear in a clearly recognisable form until about the 9th century AD. Furthermore, certain early plagues such as the plague of Athens of 429 BC, which is vividly detailed by the Greek historian Thucydides, do not seem to have an easily recognisable modern counterpart.

We noticed that epidemics and pandemics of fresh diseases, both in historical times as well as more recently, have almost without exception appeared suddenly and spread with phenomenal swiftness. The influenza pandemics of 1889–1890 and 1918–1919 both swept over vast areas of the globe in a matter of weeks. Such swiftness of spread, particularly in days prior to air travel, is difficult to understand if infection can pass only from person to person. Rather it is strongly suggestive of an extraterrestrial invasion over a global scale. We argued now that it is the primary cometary dust infection that is the most lethal, and that secondary person-to-person transmissions have a progressively reduced virulence, so resulting in a diminishing incidence of disease over a limited timescale.

On this picture, the pattern of incidence and propagation of any particular invasion is a somewhat complex matter. It depends, amongst other factors, upon the sizes of incoming micrometeoroids, the local physical characteristics of the atmosphere and on the distribution of global air currents. We might expect certain latitude belts on the Earth to be comparatively disease free, whilst others might be more prone to receiving space-borne pathogens. Also, depending on sizes of particles amongst other factors, some epidemics may be geographically localised while others may be global. In all cases any new epidemic must occur suddenly — when the Earth crosses the trail of infected cometary particles.

Fred and I became fully convinced that these ideas had to be substantively right. We both made a great effort to learn as much as we could about virology and infectious diseases, from text books as well as by talking to our medical colleagues at the University Hospital, particularly John Watkins, Professor of Medical Microbiology and Robert Mahler, Professor of Medicine. Fred's visits to Cardiff were now taken up with locating the right virologists, bacteriologists and historians who could enlighten us with facts about diseases past and present. We also made several visits together to the Central Virus Reference Laboratory in Colindale, London, and one memorable visit to meet Sir Christopher Andrewes (a virologist who played a role in isolating cold-type viruses) at the Common Cold Research Centre in Salisbury plains. Here we discovered that all attempts to infect volunteers with a common cold virus under controlled, epidemic-like conditions had been, up to that time, a failure.

We soon began to take a particular interest in one disease — influenza — because we discovered there were many puzzling aspects connected with its epidemiology (epidemic behaviour). Here was a disease that appeared to indicate the incidence of a virus or at least a trigger from the atmosphere. A distinguished epidemiologist Charles Creighton maintained as late as the final decade of the 19th century that influenza is not a transmissible disease. In his book *History of Epidemics in Britain* (Cambridge University Press, 1891) he discusses the influenza epidemics of 1833, 1837 and 1847, in which medical opinion held that populations living over considerable areas are affected almost simultaneously. Such evidence suggested to Creighton a "miasma" descending over the land rather than a disease which must spread itself from person to person. If one substitutes for "miasma" the phrase "viral invasion from space" it is a similar position to that which we arrived at in 1977. Creighton's hegemony, however, was short lived, and by the end of the 19th century, the concept of infectious disease caused by Earth-bound microorganisms had firmly taken root.

Strictly, the microbiological concept requires only that victims of the disease should acquire the virus from outside of themselves, which of course they would do if the infection came from the atmosphere. But such an idea seemed so much less plausible to scientific opinion than the concept of person-to-person transmission — this was not even considered as a hypothesis to be tested. It became an axiom.

Fred Hoyle and I became convinced that the transmission hypothesis could be tested if a new pandemic strain arose. Particles of viral size added to the Earth are stopped in the atmosphere at a height of about 30 km. Vertical descent of particles through the stratosphere and into the troposphere occurs mainly as a result of winter downdrafts that occur 6-months apart in the two hemispheres. This phenomenon offers a ready explanation of the fact that influenza and other respiratory viral diseases are distinctly seasonal in character.

It is commonly found that lingering mists in the winter season usher in a wave of flu-like disease. Since, as I have already said, bacteria and viruses can act as condensation nuclei around which water droplets form, this coincidence is not entirely unexpected. In situations where rain falls as large drops there is not much chance of direct inhalation of nucleating viruses, whereas misty weather provides the incoming virus with the best opportunity to become dispersed in aerosol form in a way that can easily be inhaled near ground level.

Among earlier evidence that pointed in this direction, the observations of Professor Magrassi in 1948 are worthy of note. The worldwide influenza epidemic of 1948 apparently first appeared in Sardinia. Magrassi, commenting on this epidemic, wrote:

> "*We were able to verify the appearance of influenza in shepherd who were living for a long time alone, in open country, far from any inhabited centre; this occurred almost contemporaneously with the appearance of influenza in the nearest inhabited centres...*".

One of the most striking features in this whole story is that the technology of human travel has had no effect whatsoever on the way that influenza spreads. If influenza is indeed spread by contact between people, one would expect the advent of air travel to have heralded great changes in the way the disease spreads across the world. Yet the spread of influenza in 1918, before air travel, was no faster, and no different from its spread in more recent times.

Probably the most disastrous influenza pandemic in recent history occurred in 1918–1919 and caused about 30 million deaths. After studying all the available information about the spread of influenza during this epidemic Dr. Louis Weinstein wrote:

> *"Although person-to-person spread occurred in local areas, the disease appeared on the same day in widely separated parts of the world on the one hand, but on the other hand, took days to weeks to spread relatively short distances. It was detected in Boston and Bombay on the same day, but took three weeks before it reached New York City, despite the fact that there was considerable travel between the two cities. It was present for the first time at Joliet in the State of Illinois four weeks after it was first detected in Chicago, the distance between those areas being only 38 miles..."*

As we were pondering on such matters serendipity intervened once again with a remarkable circumstance. The new pandemic we had talked about had become a reality. In November 1977 an outbreak of a strain of flu that had not been present in the human population for 20 years, was reported in the Far East region of the old Soviet Union. By the end of December the first cases were reported in Britain and by January the disease was rampant, most noticeably it was raging through the schools of England and Wales.

Our ideas on the emergence of life on comets and of a possible connection with plagues and pestilences had now advanced to a stage when I felt that Fred should try them out on an academic audience. Fred obligingly agreed to deliver a lecture entitled "Diseases from Outer Space" at Cardiff on January 18, 1978. Needless to say his lecture was a sell-out. There was only standing room in the main auditorium of Cardiff's largest lecture theatre at the time, the Sherman Theatre. His lecture, chaired incidentally by Principal Bevan, was greatly appreciated even though it stimulated controversy and a degree of hostility as well. The so-called Red Flu was around us everywhere, as was evident even in the coughing that was heard through Fred's lecture. During the days he spent in Cardiff, Fred and I developed a strategy to investigate this epidemic.

We saw this as an ideal opportunity for testing person-to-person transmission. School children under the age of 20 had not been exposed to the new virus in their lifetime, and so were all equally susceptible. We had the idea of using the school population as detectors of the new virus, rather in the same way that physicists use amplifying detectors to observe small fluxes of incident cosmic rays. Our first "experiment" was confined to schools, including boarding schools in South Wales and the South West region of England. We began with analyses of school absenteeism. For this purpose Priya and I did a mammoth circuit of schools within 30 miles of our house, and examined their attendance records. Our objective was to determine, for each individual school, the time dependence of absenteeism due to influenza during the epidemic, and also the overall attack rates that could be inferred. What surprised us most was the huge range in the attack rates, essentially 0% to over 80%, and this was determined only by the location of the school, indicating that the incidence of the virus was patchy on distance scales of a few kilometres or less. Data that Priya and I had collected for boarders in Howells School Llandaff, Cardiff and Atlantic College at St Donats gave a taste of more to come. There was clearly an effect connected with the houses where the children slept; some had high attack rates, others very low. Already, person-to-person transmission was beginning to look unlikely.

Fred and I published our preliminary findings in *New Scientist* (September 28, 1978) and started a more ambitious exploration of the school data throughout England and Wales. We circulated a questionnaire to all privately supported secondary schools seeking the following information:

1. Overall attack rates amongst boarders and day pupils (separate data) during the recent influenza epidemic;

2. Day-to-day attack patterns as they are reflected in classroom absence and/or admissions to school sick bays;
3. The date of the peak of epidemic experienced in each school.

The results of our bigger survey only confirmed and strengthened those we had from the local schools. It is commonly stated that school boarders, members of the armed services in stations, and other closed communities, are highly susceptible to epidemic outbreaks of influenza. Replies to our questionnaire showed clearly that as far as school boarders are considered this is a myth. Our sample involved a total of more than 20,000 pupils with a total number of victims of some 8800. The distribution of attack rates in the schools showed that only three schools out of more than a hundred had the very high attack rates that have been claimed to be the norm.

If the virus responsible for the 8800 cases were passed from pupil to pupil, much more uniformity of behaviour would have been expected. We found evidence for great diversity, with a hint that the attack rate experienced by a particular school (or house within a school) depended on where it was located in relation to a general infall pattern of the virus. The details of this infall pattern are determined by local meteorological factors. The infall clearly displayed patchiness over a scale of tens of kilometers, the typical separation between the schools.

One particular school in our survey, Eton College, merits special attention. There were 1248 pupils distributed in a number of boarding houses and the total number of cases across the whole school was 441. The actual distribution of cases by house showed enormous heterogeneity. College House with a total population of 70 had only one case, compared with the expected value of 25 on the assumption of random distribution, in a person-to-person infection model. Here again we saw heterogeneity, but now on the scale of hundreds of metres. This entire distribution would be expected once in 10^{16} trials on the basis of person-to-person transmission. Clearly, if one looks objectively at all the facts, flu cannot be "catching" from person to person as our present-day scientific culture would have us believe.

Further evidence against the standard dogmas of influenza transmission came from a study of influenza in Japan. The Japanese data was supplied by my friend Shin Yabushita, a contemporary of mine at Cambridge. Japan is remarkable in that several large areas have population densities in excess of 2000 per square km, whereas others have well under 200 per square km. Standards of monitoring and reporting disease are also uniformly efficient throughout most of Japan. The data showed extreme variability of attack rates between adjacent prefectures. Here again the patchiness of viral infall is over a distance scale of tens of kilometres, exactly as in the case of the Welsh and English schools we already saw.

We have already noted that the descent of the virus from the stratosphere to ground level depends on global circulation patterns of the atmosphere. This fact accounts for the otherwise mysterious phenomenon that epidemics of flu occur with a distinct seasonality in widely separated parts of the world.

Scientists who feel uncomfortable with the logical inferences drawn from a theory such as this often seek eye-catching one-line disproofs. For the case of influenza from space the often-stated disproof is that: "Viruses are host-specific, and so must have evolved in close proximity to the terrestrial species that they attack". In other words the critic says: "How could the incoming virus know ahead of its arrival the nature of the highly-evolved and specialised hosts that it may encounter?" Our answer is simple: "The virus could not of course anticipate us, but we, the host species, could anticipate the virus since we must have had a long and continuing exposure to viruses of a similar kind". It is also well known that viruses can on occasion add onto our genes and so viral DNA sequences serve as an invaluable store of evolutionary potential. Our genomes would be "made up" of cometary viruses according to this point of view. If the influenza virus, or one that is similar to it, forms part of our genetic heritage, then the so-called host specificity, or the apparent human-virus connection is instantly and elegantly explained.

According to our point of view, reservoirs of the causative agent for influenza are periodically resupplied at the very top of the Earth's atmosphere. Small particles, the sizes of viruses or smaller, tend to remain suspended high up in this region for long periods unless they are pulled down into the lower atmosphere. In high latitude countries, such breakthrough processes, where the upper and lower air becomes mixed, are seasonal and occur during the winter months. Thus a typical influenza season in a European country would occur between December and March. Frontal conditions with high wind, snow and rain effectively pull down viral pathogens close to ground level. The complex turbulence patterns of the lower air ultimately control the details of the attack at ground level, and determine why people at one place and at one time succumb, and why those in other places and at other times do not.

We also discovered in the literature that ozone measurements can be used to trace the mass movements of air in the stratosphere. Such measurements show a winter downdraft that is strongest over the latitude range from 40° to 60°. Taking advantage of this annual downdraft, individual viral particles incident on the atmosphere from space would therefore reach ground-level generally in temperate latitudes. Such locations on the globe would naturally emerge as places where upper respiratory infections are likely to be most prevalent, on the supposition of course that the Earth is smooth. The exceptionally high mountains of the Himalayas, rearing up through most of the height range to the stratosphere, introduce a large perturbation on the smooth condition, which may be expected to affect adversely this particular region of the Earth, especially regions lying downwind of the Himalayas, particularly China and South East Asia. In effect, the Himalayas are so high that they could act as a drain plug for most of the viruses incident on the atmosphere at latitude ~30°N, the large population of China being inundated by this drainage effect, making China the quickest and worst affected region of the Earth. This could explain why new respiratory viruses such as Severe Acute Respiratory Syndrome (SARS) and new influenza viruses often make their first appearance in China. Concomitantly, other parts of the Earth at ~30°N should be largely free of viral particles, unless it happens that such particles are incident as components within larger particles which fall fast under gravity.

A direct demonstration that the general winter downdraft in the stratosphere occurs strongly over the latitude range 40° to 60° was given by M.I. Kalkstein (*Science* **137**, 645, 1962) in the last of the series of atmospheric nuclear tests carried out in the middle of the 20th century. A radioactive tracer, Rh-102, was introduced into the atmosphere at a height above 100 km and the fall-out of the tracer was then measured year by year through airplane and balloon flights at altitudes ~20 km. The tracer was found to take about a decade to clear itself through repeated winter downdrafts, and this happened mostly over the latitude belt 40–60 degrees.

The observed fallout patterns of a radioactive tracer agreed closely with the well-known winter season of the viruses responsible for the majority of upper respiratory infections in temperate latitudes. The time of a decade or so that was taken to clear the tracer from the stratosphere also coincides with the average time of prevalence of any new influenza subtypes after it is first introduced. How could all these facts be explained by the conventional ideas about influenza? Fred Hoyle answered this question in his own inimitable style in one of the preprints we put out at the time:

> "*Unfortunately so little has been understood of the mode of attack of the so-called infectious diseases that almost any form of hypothesis has come to be accepted in the past as an answer to questions of this sort. The truth is that, although the world may be extremely complex it is nevertheless extremely precise, with explanations every bit as clear-cut as that of the quantum mechanical analysis of the energy levels of the hydrogen atom being ultimately available for every phenomenon we observe*".

If one looks at the disc of the sun through specialised darkened glass dark spots are often to be seen. These are vortices of gas on the surface that are associated with strong magnetic fields and their numbers vary enormously through a solar cycle which lasts about 11 years. I have referred earlier to my first scientific paper with Fred in 1962 proposing a mechanism for periodic reversals of the sun's polar magnetic field. This reversal is connected with the cumulative effects of such sunspot activity. Sunspot numbers give a measure of high-energy activity at the sun's surface, the peak numbers corresponding with frequent solar flares and the emissions of charged particles that reach the Earth. Such activity on the sun is known to result in geomagnetic storms, ionospheric disturbances that interfere with radio communications, and most spectacularly the production of bright auroral displays (Northern Lights), the latter being caused by the streaming of charged particles from the sun moving along magnetic field lines.

A possible connection between peaks of sunspot activity and the times of influenza epidemics was first suggested by Edgar Hope-Simpson on the basis of data over a limited time span. We extended these comparisons over a longer time interval and found that a correlation of this kind did indeed hold. Peaks of solar activity will undoubtedly assist in the descent of charged molecular aggregates (including viruses) from the stratosphere to ground level. Thus according to our point of view serious influenza epidemics would follow such peaks, provided the culprit molecular aggregates were recently dispersed in the stratosphere from cometary meteor streams. With a more or less regular occurrence of such meteor showers, the limiting factor may be the intensity of solar activity, leading to coincidences between the timings of pandemics or major epidemics and sunspot peaks.

With all these considerations well in place on influenza as well as other epidemic diseases we now began to write our second book *Diseases from Space*, published in 1979 by J.M. Dent.

1. My father, who inspired me to start my journey, outside the Senate House, Cambridge, 1932.

2. With my brothers *en route* to England on a visit, 1946. Dayal, who figures in the story, is on the left.

3. Sunset at a beach near my home in Colombo, 1960.

4. With a group of Commonwealth Scholars on arrival in the UK, outside Houses of Parliament, 1960. (I am on the left.)

5. With Jayant Narlikar at Granchester, Cambs, 1961.

6. My trip to the Lake District with Fred, 1961.

7. Fred looks up to muse on the rain clouds, 1961.

8. Fred in animated conversation at a Varenna Summer School in 1961.

9. Mayo Greenberg has a captive audience in Troy, NY, 1965.

10. Enter Priya, Sri Lanka, 1966.

11. Dining on board SS Oriana, 1966.

12. Barbara Hoyle and Priya stop to drink the sap of a coconut, during Barbara's visit to Sri Lanka, 1969.

13. Barbara Hoyle, Viv Howes and Priya at "Land's End" Sri Lanka, 1969.

14. A family group with the Hoyles at Gregynog, 1977. *Left to Right*: Anil (my son), Liz (Fred's daughter), Fred, Geoff (Fred's son), Samantha (Fred's granddaughter), Nina Solomon, Jacqueline (Fred's granddaughter) and Kamala (my elder daughter).

15. My children and Fred's grandchildren with Matsuda. (Courtesy Anna Jones).

18. Priya signing copies of *Spicy and Delicious*, 1979.

16. Discussion at the black board, 1977.

19. With Fred and Bill Bevan at the launching of University College Cardiff Press, 1980.

17. With Fred in my office, Cardiff, 1978.

20. Fred and Bevan: Questions after a Public Lecture, 1980.

21. At the Arkansas Trial, with Priya and infant Janaki, 1982.

22. Presenting a copy of *Lifecloud* to President J. R. Jayawardene, President of Sri Lanka, 1989.

23. The Dyffryn Gardens Meeting on Alternative Cosmologies in 1989. *Left to Right*: Jayant, Geoff Burbidge, Fred, Chip Arp, and myself.

24. Discussions with Arthur C. Clarke in Sri Lanka, c 1995.

25. Jayant, Fred and Fred in bronze cast by Shiela Solomon, c 1997.

26. Fred looking again at *Space Travellers*, April, 2001.

Chapter 12

First Signs of Life

I had wondered for a while what astronomical discovery would encourage Fred to take the step from life emerging on a comet out of interstellar prebiotics to fully-fledged microbial life distributed throughout interstellar space. He had now been expounding the former thesis for nearly two years with conviction and much eloquence, and had he stayed with that our fortunes may have turned out differently. Our position of 1977 was after all the standard point of view of the scientific community in 2004, and we may now have been more openly acknowledged as its pioneers.

As a brief respite from our work on the influenza pandemic I decided upon impulse to take a closer look at the visual extinction curve of starlight, the way that starlight is extinguished by dust at visual wavelengths. This, it would be recalled, is where my research into this whole subject began in 1961. There were many unresolved problems that had to be solved. Over the wavelength range from 7000 to 3000 Å, the extinction (dimming ratio on a logarithmic scale) was approximately inversely proportional to the wavelength, and this was the case in whatever direction one looked. Such an invariance of behaviour was difficult to reconcile with the grain models we had discussed so far involving mixtures of silicate or organic grains along with graphite and iron. In all these non-biological grain models the visual part of the extinction curve had to come mostly from the dielectric (non-absorbing) component of the mixture, and to get the correct shape of the extinction curve, grain radii had to be fairly sharply fixed. This condition could not be relaxed as long as one

109

stayed with grain materials such as ice, organics or silicates which have visual refractive index values $n = 1.3, 1.5, 1.6$, respectively. In other words, the solutions obtained so far in all these cases are highly parameter sensitive, and therefore not very satisfactory.

I discovered that one way to relax the size constraint is to reduce the value of the refractive index, n, below 1.3. A value closer to $n = 1.15$ would be nearly optimal from this point of view, for the reason that the extinction efficiency of spheres with this refractive index can remain closely proportional to inverse wavelength over a wide range of wavelengths. But what material could possess such a low value of n? When I searched the relevant handbooks of physical constants it soon became obvious that, with the exception of solid hydrogen, there was no other homogeneous solid material that had the desired property. Solid hydrogen was of course ruled out because we had found earlier that it could not survive under normal interstellar conditions. It was at this point I recalled a collaboration I had been engaged in with Craig F. Bohren (C.F. Bohren and N.C. Wickramasinghe, *Astrophys. Space Sci.* **50**, 461–472, 1977) on the optical properties of heterogeneous grains, that is to say grains comprised of a mixture of two different particulate types welded together. I immediately turned to this earlier work, noting that a particle made of organic material with bulk refractive index $n = 1.5$ could have an average refractive index of 1.2 or less if it contained minute vacuum cavities. For instance an aerogel would have such a refractive index.

When I pointed this out to Fred his interest in the possibility of bacterial grains in space was immediately aroused. We got to work on discussing ways in which bacteria would freeze-dry in space leading to the production of vacuum cavities within them. We started with a vegetative bacterial cell with the following typical constitution: Organic material 20%, bound water 20%, free water 60%. Freeze-drying in a vacuum, such as in outer space, would maintain the cell wall intact and also retain the interior organic content and bound water, while free water will escape and lead to the production of vacuum cavities. The average refractive index (by a straightforward averaging process) could then be shown to be

$$n = 0.2 \times 1.5 + 0.2 \times 1.3 + 0.6 \times 1.0 = 1.16.$$

For freeze-dried interstellar bacteria we concluded that it would be reasonable to assume an average refractive index in the range 1.15–1.16.

The next thing we needed to know was the size distribution that would be representative for actual bacteria; and for this, *Bergey's Manual of Determinative Bacteriology* was consulted. The best data we could find related to spore-forming bacteria, which gave the count of species in various ranges of sizes in the form of a histogram. By taking freeze-dried interstellar bacteria to have this particular size distribution that was appropriate for terrestrial spore forming bacteria, we had a situation totally different from anything we had experienced previously. For now there were no parameters at all to be fitted. The extinction behaviour of the entire ensemble of grains became immediately amenable to calculation using the computer program I had developed and used over many years. The result was staggering: we had discovered a perfect fit to the average interstellar extinction over the visual waveband with just the one assumption *that interstellar grains were freeze-dried bacteria.* Figure 7 shows this fit

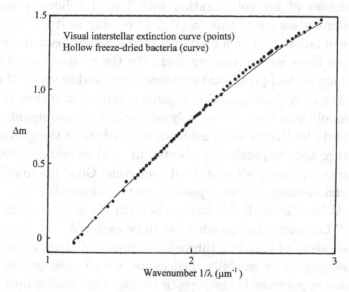

Fig. 7   Interstellar extinction observations over the visible waveband (points) compared with the theoretical extinction curve calculated for hollow freeze dried bacteria (curve).

and the solution, it should be stressed, is essentially parameter free. Referring to such matters in a public lecture Fred said:

"I have a particularly ferocious dislike of model calculations which achieve a tolerable correspondence with observation through assigning more-or-less arbitrary values to many parameters introduced by the investigator. Although in the past I strayed from the strict and narrow path of virtue by engaging in this self-deceiving practice, nowadays I refuse to spend five minutes on such parameter-fitting exercises. By a like token I am heavily impressed by agreements achieved by models without arbitrary parameters, and I am doubly impressed by agreements in which such calculations precede in time what the observations eventually turn out to be. In short, successful model calculations without free parameters that are predictive are for me doubly impressive."

Fred's adherence to this philosophy was evident throughout the four decades of my collaboration with him. I believe it was the extraordinary nature of this fit (Fig. 7) coming as it did after two · decades of failure that won Fred over to the case for bacterial grains. And then there was no turning back. For the composition of interstellar dust we had progressed cautiously and in slow stages through a sequence of options: graphite, organic polymers, and then to complex biopolymers such as the polysaccharides. These organic polymeric particles that were in evidence everywhere in the galaxy had an average size comparable to a bacterium, and an average refractive index appropriate for a freeze-dried bacterium. Good fits to infrared, visual and infrared data were possible on the assumption of bacteria-like particles. Could all this be somehow explained without invoking biology? Of course this question has to be explored.

After weeks of fumbling through a sequence of ideas, all of which were proving to be woefully inadequate, we alighted on the most promising hypothesis. In the gigantic clouds of interstellar dust could we be witnessing no less than the dissemination of biology itself? Could the interstellar medium be choc-a-bloc, not simply with the

building blocks of life arranged into prebiotic molecules or even pro-
tocells, but with the end products of the living process itself? And
this would then be required to happen on an unimaginably vast scale.
At the end of a long run of frenzied telephone calls between Cardiff
and Cockley Moor we decided that was it! Interstellar grains must
surely be bacteria — albeit freeze-dried, perhaps mostly dead! At the
very least this was a hypothesis that had to be explored.

Nobody takes lightly the prospect of walking into exile, albeit
scientific exile, and moreover one that is self-imposed. Yet by the
Spring of 1977 it appeared to both of us that we had no option but
to do so, carrying the heavy burden of not one but two scientific
heresies: the heresy of diseases from space and now the heresy of
microbial life in interstellar space. No sooner than we discovered the
solution in Fig. 7 (which is further elaborated in Fig. 8) we wrote
up our results, distributed them to colleagues in the form of a Blue
Cardiff Preprint at the beginning of April 1979, and at the same
time sent off a technical paper entitled "On the nature of interstellar
grains" for publication in *Astrophysics and Space Science* (Hoyle and
Wickramasinghe, 1979b).

The immense power of bacterial replication is worth careful note
at this point. Given appropriate conditions for replication, a typical
doubling time for bacteria would be two to three hours. Continuing
to supply nutrients, a single initial bacterium would generate some
2^{40} offspring in 4 days, yielding a culture with the size of a cube
of sugar. Continuing for a further 4 days and the culture, now containing 2^{80} bacteria, would have the size of a village pond. Another
4 days and the resulting 2^{120} bacteria would have the scale of the
Pacific Ocean. Yet another 4 days and the 2^{160} bacteria would be
comparable in mass to a molecular cloud like the Orion Nebula. And
4 days more still for a total time since the beginning of 20 days, and
the bacterial mass would be that of a million galaxies.[1] No abiotic
process remotely matches this *potential* replication power of a biological template. Once the immense quantity of organic material in

---

[1] As the readers would know, this is not possible since a "continuous supply of
nutrients" to such a massive colony is inconceivable!

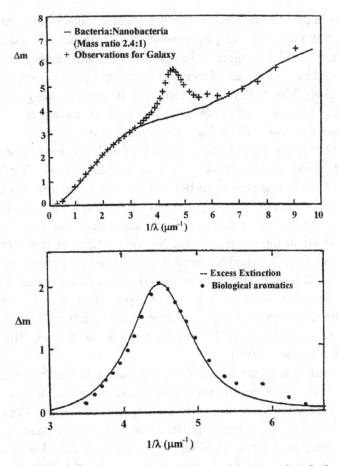

Fig. 8   Interstellar extinction observations (crosses) compared with the scatter-ing behaviour of freeze-dried bacteria and nanobacteria (curve, upper panel). Lower panel shows how the excess extinction over the scattering background in the 3300 Å–1500 Å waveband is matched by the properties of biological aromatic molecules.

the interstellar material has been appreciated, a biological origin for it becomes a necessary conclusion.

But where are astronomical locations where conditions for repli-cation of bacteria can be found? Certainly not in the cold depths of space, where microbes could merely remain in a freeze-dried dormant state. Planets like the Earth would provide too small a total mass of carbonaceous material to make any cosmic impact. It is therefore to comets we turned, arguing that comets are the sources of biological particles in interstellar clouds. An individual comet is a rather insubstantial object. But our solar system possesses so many of them, perhaps more than a hundred billion of them, that in total mass they equal the combined masses of the outer planets Uranus and Neptune, about 10^{29} grams. If all the dwarf stars in our galaxy are similarly endowed with comets, then the total mass of all the comets in our galaxy, with its 10^{11} dwarf stars, turns out to be some 10^{40} grams, which is just the amount of all the interstellar organic particles that are present in the dust clouds within the galaxy.

How would microorganisms be generated within comets, and then how could they get out of comets? We know as a matter of fact that comets do eject organic particles, typically at a rate of a million or more tons a day when they visit the inner regions of the solar system. We argued that comets when they are formed incorporate interstellar bacterial particles, from which only a fraction 10^{-22} needs to retain viability for a regeneration process to operate. For at least a million years, at the time of their origin, comets have liquid cores due to radioactive heat sources such as ^{26}Al which are also incorporated within them. Within a very brief period as described above sequential doublings of viable microorganisms would lead to a sizable portion of the cometary core being converted into biomaterial. When the comets re-freeze this amplified microbial material is also frozen in, only to be released when they become periodically warmed up in the inner solar system. Some of this bacterial matter may reach the inner planets, which they can seed with life, some of it is expelled back into interstellar space.

Our point of view requires that bacteria must be space-hardy which recent research has shown is the case. On the whole microbiological research of the past 20 years has shown that bacteria and other microorganisms are indeed remarkably space-hardy (J. Postgate, *The Outer Reaches of Life*, Cambridge University Press, 1994). Microorganisms known as thermophiles and hyperthermophiles are present at temperatures above boiling point in oceanic thermal vents. Entire ecologies of microorganisms are present in the frozen wastes of the Antarctic ices. A formidable total mass of microbes exists in

the depths of the Earth's crust, some 8 kilometres below the surface, greater than the biomass at the surface (T. Gold, *Proc. Natl.
Acad. Sci.* **89**, 6045–6049, 1992). A species of phototropic sulfur bacterium has been recently recovered from the Black Sea that can
perform photosynthesis at exceedingly low light levels, approach-
ing near total darkness (J. Overmann, H. Cyoionka and N. Pfennig,
Limnol. Oceanogr. **33**(1), 150–155, 1992). There are bacteria (e.g.
Deinococcus radiodurans) that thrive in the cores of nuclear reactors.
Such bacteria perform the amazing feat of using an enzyme system
to repair DNA damage, in cases where it is estimated that the DNA
experienced as many as a million breaks in its helical structure.

There is scarcely any set of conditions prevailing on Earth, no
matter how extreme, that is incapable of harbouring some type of
microbial life. As for ultraviolet damage under space conditions, this
is very easily shielded against. A carbonaceous coating of only a few
microns thick provides essentially a total shielding against ultraviolet light; and there are several modern experiments that have
demonstrated precisely that. Next, let us note that many types of
microorganisms are not really killed by ultraviolet light; they are only
deactivated. And this happens through a shifting of certain chemical
bonds contained in the genetic structures of the organisms, without
destroying the genetic arrangements themselves. And this permits
the original properties to be recovered once the ultraviolet radiation
has been shut off. Furthermore, we know that microorganisms that
are normally sensitive to ultraviolet light can, through repeated expo-
sures, be made just as insensitive as the more resistant kinds — yet
another unearthly property.

All this was good news for our theory, but as with every good
theory we had to be able to demonstrate its predictive capabilities.
We had to make predictions that could be verified or falsified by
future experiment or observation. This was yet a couple of years
in the future. In the years 1979 and 1980 when these crucial steps
were taken in our collaboration, the world outside was recording a
few dramatic events. The Shah of Iran was driven into exile; there
were referenda for home rule in Scotland and Wales, the Scots say
yes, the Welsh say no; Margaret Thatcher becomes Britain's first

woman Prime Minister. More relevant to our story the US spacecraft Voyager I returns the most dramatic pictures of Jupiter, Saturn and Uranus. Rings are discovered around both Jupiter and Uranus, leading us, Fred and myself, to speculate whether these contained bacterial particles.

On October 29th–31st 1980, a conference with the title "Comets and the Origin of Life" was organized by Cyril Ponnamperuma at the University of Maryland. Ponnamperuma, a Sri Lankan born chemist, always had his feet firmly planted in the opposite camp from us. He had made important laboratory studies on the abiotic synthesis of organic molecules, including sugars and nucleotide bases under conditions similar to those used in the famous Urey–Miller experiment of the 1950's. All such experiments are of course a very far cry from the generation of life, but they are often presented as being a significant step in that direction. Ponnamperuma was at the time Director of the Laboratory of Chemical Evolution at the University of Maryland, and clearly was an opponent of the ideas being discussed by Fred and myself. I was, however, invited by him to present a joint paper with Fred entitled "Comets — a vehicle for panspermia," an invitation that I accepted with some trepidation.

Cyril Ponnamperuma had rounded up just about all of our potential adversaries, including J. Oro and of course J. Mayo Greenberg. Almost everyone at the meeting delivered a polemic denying any possibility of comets carrying life, or even life's building blocks in some instances. Greenberg made his first coherent claim to rival our 1974 ideas of polymeric grains with his own title "Chemical Evolution of Interstellar Dust — A source of prebiotic molecules". He was unfortunately a few years too late to steal any priority from Fred and myself. The conference volume, including our paper, was published by D. Reidel Co. in 1981 (*Comets and the Origin of Life* — ed. C. Ponnamperuma). Nothing that I heard at this meeting made me weaken my resolve to continue our search for an origin of life on a grand cosmic scale.

Chapter 13

Bacterial Dust Predictions Verified

The conference in Maryland gave me the first direct experience of the power of the opposition that was rallied against us. I formed the impression that nothing will be spared in an attempt to denigrate our work or to stifle its further progress. The hostility was getting steadily worse as the evidence grew in strength. Fred put this cogently in a piece he wrote for our joint book *From Grains to Bacteria* (University College Press, 1984):

> *"It is necessary to come now to a curious situation that we think will eventually be of interest to students of scientific methodology. The more precise the correspondences we calculated between our models and the observation, the greater was the measure of opposition we received from individuals, from journals and from funding agencies like SERC. The introduction of polysaccharides, because of their biological association apparently, became a signal for papers to be turned down by journals, and even for the most modest grant applications to be thrown back in our faces by SERC, an organisation which in a time span of no more than a decade and a half managed to go from a beginning of rich promise to one of the outstanding Gilbert-and-Sullivan operattas of the twentieth century ...".*

Such remarks, coming from one who was not so long ago Chairman of a SERC committee, had a sense of cynicism as well as disillusionment. It was during the period 1979–1981 that we encountered the worst of the opposition that Fred describes. This entailed a succession of socially welcomed one line disproofs that came to be offered. For example it was often stated that organic matter could be expected to have many C–H linkages, with absorption due to stretching in the 3.3–3.5 μm wavelength range in the infrared region of the spectrum. Claims were published in refereed journals that this absorption had been looked for and had not been found. Therefore, it was argued, interstellar grains could not be of an organic nature as we had claimed. This was an argument which those with an anti-Copernican opposition to the thought that life might be cosmic may have found appealing, but unfortunately it turned out to be false.

The following quote from W.W. Duley and D.A. Williams published in *Nature* (**277**, 4 January, 1979) (one of many similar quotes) illustrates this state of mind:

> "*We conclude that no spectroscopic evidence exists to support the contention that much of interstellar dust consists of organic materials. While the presence of trace quantities of organic compounds on grains inside very dense clouds cannot be excluded by available data, the absence of any observation of a 3.3–3.4 μm absorption band even in objects with* $A_v = 50$ *mag strongly suggests that organic grains constitute at most a minor component of interstellar dust...*"

How wrong this has been because, in 2004, there is not the slightest dissent from the view that the bulk of interstellar dust is organic. The illusion for some in 1979 came from neglecting to take proper account of the band strength of the C–H linkages in actual biopolymers which we knew from the measurements of Tony Olavesen in Cardiff to be weak, with a mass absorption coefficient of not more than about $1000 \, \text{cm}^2 \, \text{g}^{-1}$. That is to say, the expected absorption in the 3.3–3.5 μm band could not exceed about 2% of the visual extinction. Such a small effect was not detectable even for the most highly reddened stars with the equipment and instruments available in 1979, so the claims made by David Williams and others were incorrect.

We had already shown that the C–H stretching organic band was present as a minor shoulder in the wings of 3 μm bands in galactic infrared sources, and the fits of polysaccharide models to such sources were exceedingly close. Also the extinction curve of starlight for a biological model matched the astronomical data with uncanny precision.

With the emergence of Cardiff as a new centre for astronomy, the Royal Astronomical Society (RAS) asked me to arrange their annual out of town meeting for 1980 in Cardiff. The then RAS President, Professor M.J. Seaton of University College London also requested Fred Hoyle to give an evening public lecture on this occasion. On April 15, 1980 Fred delivered his lecture with the title "The Relation of Biology to Astronomy" to a saturated hall at University College Cardiff, in which he presented an eloquent exposition of our position on the nature of interstellar grains. His frontal assault on conventional theories of biological evolution on the Earth did not win him much support, but the case against such theories was outspokenly presented. For instance:

> *"What may be the biggest biological myth of all holds that evolution by natural selection explains the origin of the phyla, classes and orders of plants and animals. There are certainly plenty of examples of minor evolutionary changes caused by natural selection, and on the evidence of these minor changes the major changes are assumed to be similarly caused. The assumption became dogma, and then in many people's eyes the dogma became fact..."*

This was a statement of intent that we were soon to take on the entire biological establishment in re-evaluating the evolution of life in a closed box setting, and opening up the process to the wider universe. This was to happen in the months that followed leading eventually to the publication of a book (*Evolution from Space*) and several short booklets.

I suspect that the majority of the audience in Cardiff, who were astronomers, did not take much interest in biology at this stage even though it was the dominion of astronomy that was being re-evaluated and enlarged. Thus he went on:

> *"Astronomers have become accustomed to thinking of the external Universe in the words of Macbeth, as being 'full of sound and fury, signifying nothing'. Can we seriously believe*

that anything as subtle as biology could have gained a toe-hold in a world signifying nothing? I pondered this question for a long time before arriving at a strange answer to it. If the astronomer's world of fury is really in control, then the prospects for biology would be poor. But what if it is really biology which controls the astronomer's world?"

Unwittingly perhaps Fred laid the foundations for the modern discipline of astrobiology, a subject that is becoming increasingly popular these days, with his concluding remarks on April 15, 1980:

"Microbiology may be said to have had its beginnings in the nineteen-forties. A new world of the most astonishing complexity began then to be revealed. In retrospect I find it remarkable that microbiologists did not at once recognise that the world into which they had penetrated had of necessity to be of a cosmic order. I suspect that the cosmic quality of microbiology will seem as obvious to future generations as the Sun being the centre of our solar system seems obvious to the present generation."

In the days immediately following Fred stayed with us as usual. We took this opportunity to identify many of the loose ends of our theory that had to be dealt with. The immediate question now was: what further checks were there to be made for the thesis of interstellar bacteria? Predictions of the behaviour of a bacterial model at infrared wavelengths should be made, and these might then be looked for in astronomy. But this required experimental work to be carried out before the relevant astronomical observations were made.

Here was our next instance of the intervention of serendipity. My brother D.T. Wickramasinghe, (Dayal), Professor of Mathematics at the Australian National University in Canberra, was also an astronomer (trained at Fred Hoyle's IOTA!) and frequently used the 3.9 metre Anglo-Australian Telescope (AAT). The AAT, the completion of which was in large measure due to Fred Hoyle's untiring efforts in the early 1970's, happened to be equipped with just the right instruments to look for a signature of interstellar bacteria.

Shortly after the release of our April 1979 preprint on the scattering properties of bacteria, Dayal visited Cardiff to spend some time with our family. Dayal's visit happened to coincide with a time when Fred Hoyle was also in Cardiff. We naturally got talking about matters relating to interstellar bacteria. Dayal asked: "What do you think can be done at the telescope to prove or disprove your theory?" to which we promptly replied that he could use the infrared spectrometers on the AAT to look at infrared sources near the wavelength of 3.4 μm in greater detail than ever before. A very long path length through the galaxy was needed to have any hope of detecting such an effect unambiguously. The longest feasible path length through interstellar dust that existed within our own galaxy was defined by the distance from the Earth to the centre of the Galaxy. There were several sources of infrared radiation located near the galactic centre that could serve as search lights for interstellar bacteria. Dayal was doubtful that he would be allocated observing time if he applied for such time specifically to do this project. The general consensus then was that life in space could not be regarded as respectable science! Dayal overcame this difficulty, however. Although honesty is the best policy, it often pays handsomely to be economical with the truth in a world of dubious morality. The deceit involved applying for telescope time to do a quite different project, and then using part of the time to look for the signature of organic matter.

In February and April 1980, Dayal, collaborating with D.A. Allen, obtained the first spectra of a source known as GC-IRS7 which showed a broad absorption feature centred at about 3.4 μm (D.T. Wickramasinghe and D.A. Allen, *Nature* **287**, 518, 1980). When Dayal's published spectrum was examined we found that it agreed in a general way with the spectrum of a bacterium that we had found in the published literature. But at this time neither the wavelength definition of the astronomical spectrum nor the laboratory bacterial spectrum was good enough to make a strong case for interstellar bacteria. However, even from this early observation we were able to check that the overwhelming bulk of interstellar dust must have a complex organic composition, in flagrant contradiction with the statements of Duley and Williams. Hitherto the infrared data showing organic polymers in space had related only to localised dust clouds such as the Trapezium nebula. More general infrared absorption by bacteria-like grains was a possibility at that point, but now it was beyond any doubt.

It was precisely at this moment that Shirwan Al-Mufti, a practical man, the son of an Iraqi Army General, approached me to become a research student at Cardiff. Here was our chance to get the required laboratory work done. We approached Tony Olavesen at the Biochemistry Department at Cardiff and arranged for Al-Mufti to be given bench space and laboratory facilities in that Department to undertake spectroscopic studies of biological samples. The purchase of a modest amount of equipment that was needed was immediately authorised by Principal Bevan, and our experimental project got under way. Al-Mufti set about executing this task with military-style efficiency, and the experiments began to yield important results in the first few months of 1981.

Al-Mufti's experiments involved desiccating bacteria, such as the common organism *E. coli*, in an oven in the absence of air and measuring, as accurately as possible, the manner in which light at infrared wavelengths is absorbed. The normal technique for doing such a measurement involved embedding the bacteria in discs of compressed potassium bromide and shining a beam of infrared light through them. The standard techniques had to be adapted only slightly. There was the need to match the interstellar environment which involved desiccation, and the spectrometer that was used had to be calibrated with greater care than a chemist would normally exercise. When all this was done it turned out that a highly specific absorption pattern emerged over the 3.3–3.6 μm wavelength region, and this pattern was found to be independent of the type of microorganism that was looked at (Fig. 9). Thus whether we looked at *E. coli* or dried yeast cells it did not matter. This invariance came as a great surprise. Our newly discovered invariant spectral signature was a property of the detailed way in which carbon and hydrogen linkages were distributed in biological systems.

The detailed structure of the pattern of absorption displayed in the upper panel of Fig. 10 is what was required to show up in

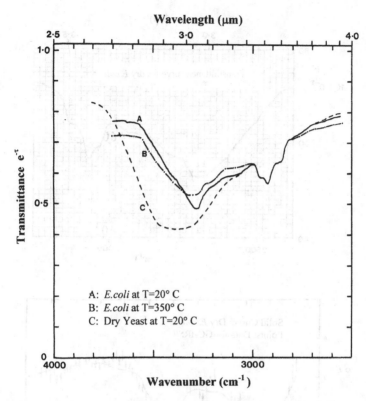

Fig. 9 Normalised transmittance properties of *E. coli* at 20 and 350 degrees Celsius and of dry yeast at 20 degrees Celsius showing an almost invariant absorption profile over the 3.3–3.7 μm waveband.

astronomy if our ideas were right. The original astronomical spectrum of GC-IRS7 of Dayal and Allen was not of high enough wavelength resolution to verify this prediction unequivocally. If astronomy turned up later with a totally different profile, then our model will have been falsified. Because we knew the exact amount of bacteria in the laboratory sample causing the absorption in Fig. 10 we could also determine how strong the absorption by bacterial dust should be at any particular wavelength. From our measurements it turned out that this absorption band was intrinsically weak. This means that a very long path length through interstellar dust was needed to get a strong positive signal confirming the presence of bacteria.

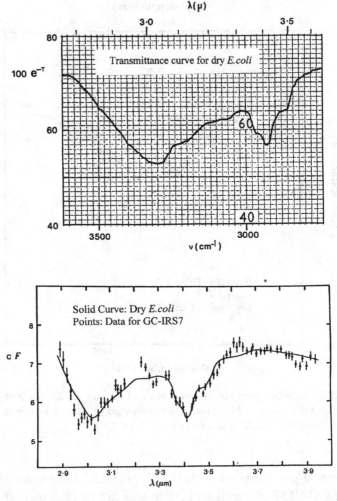

Fig. 10   Upper panel: Laboratory transmittance curve of dry *E. coli* measured by Shirwan Al-Mufti. Lower panel: Calculated behaviour of *E. coli* (curve) compared with the astronomical data for GG-IRS7 obtained by Dayal Wickramasinghe and David Allen (points with error bars) in 1981.

The observations that were to mark a crucial turning point in this entire story were carried out by Dayal and D.A. Allen at the AAT in May 1981. The new observations were of a far superior quality because a new generation of spectrometers were used. Dayal sent us his raw data by fax to compare with our new laboratory spectra which had been obtained just months earlier by Al-Mufti in March and April of the same year. After an hour or so of straightforward calculations we were able to overlay the astronomical spectrum over the detailed predictions of the bacterial model. This led to perhaps the most dramatic confirmation of the bacterial model of cosmic dust, as can be seen in the lower frame of Fig. 10. This for us was the best possible confirmation of our model, particularly because the experimental data in the comparison was obtained before the final astronomical observations became available. The agreement between a set of data points and a predicted curve as seen in Fig. 10 is normally regarded as a consistency check of the model on which the curve is based. Coming as it did after earlier fits of the same model to other sets of data, as we discussed earlier, the closeness of this particular fit would be hailed as a triumph of the model. But in our case, since the model of bacterial grains runs counter to a major paradigm in science, the situation was otherwise.

We were told by a number of chemists with experience in infrared spectroscopy that a curve like that of Fig. 10 could be obtained from non-biologically derived organic materials in many ways. Since by now we had examined without success literally hundreds of infrared spectra of organic compounds, we did not believe this claim. Consequently, we asked the chemists in question that an explicit example be produced. But it never was, with the exception perhaps that some expensive laboratory experiments, involving carefully controlled irradiation of inorganic mixtures, were claimed to yield undefined "organic residues" that may possess some of the desired properties.

We all to some degree tend to think when we run into an apparently absurd proposal, that any form of opposition to it will suffice. Because in the end what is absurd will be proved to be absurd, so that whatever we say in our opposition will eventually come out on top in the argument. Experience shows that when there are no good observations in favour of what seems absurd, this easily adopted policy is usually fairly safe. But in the face of good observations and in the face of many of them it is a highly questionable strategy.

So just how good is the agreement displayed in Fig. 10, particularly when it is taken with Fig. 7? By the early 1980's, when we attempted to answer this question, we had two decades of experience behind us in evaluating such correspondences. Expressed quite simply, we had never seen anything nearly so good. Yet even so was it all good enough to sustain a belief in such an apparently outlandish idea?

We recognised that to a person who had not followed the problem over the years the absorption characteristics of the interstellar grains in Fig. 10 might have seemed inadequate support for such a far-reaching hypothesis. And doubtlessly this was the way it appeared to many. But to us who had been involved over almost two decades it seemed otherwise, and we think it fair to add that time has supported our point of view here. Nobody among the critics of the 1980's has managed to find an alternative theory of the absorption characteristics of the grains to equal the success of the bacterial hypothesis.

And of the correspondences seen in Fig. 7 as well as Fig. 10 was not all that was there. Agreements with the data continued to emerge in whichever direction we cared to look. Our old friend the Trapezium spectrum, which was our starting point of organic dust models, now produced a perfect agreement with a biological model that combined purely carbonaceous microorganisms with a class of common algae called diatoms that included siliceous biopolymers. A microbial mix taken from a sample of water from the nearby River Taff in Cardiff resulted in the fit to the emission spectrum of the Trapezium spectrum over the entire 8–40 μm wavelength range. The correspondence is shown in Figs. 11 and 12. We referred in a previous chapter to the story surrounding the 2175 Å interstellar absorption feature and its incorrect assignment to graphite. No sooner had we embarked on the organic polymer trail than we were opting more decisively to attribute this absorption to the effect of aromatic carbon-ring structures.

From 1980 onwards infrared observations accumulated that also had a bearing on aromatic molecules (molecules involving hexagonal carbon ring structures) in interstellar space. In the mid-1980's groups of astronomers both in the USA and France independently concluded that certain infrared emission bands occuring widely in the galaxy and in extragalactic sources are due to clusters of aromatic molecules. The molecules absorb ultraviolet starlight, get heated for very brief

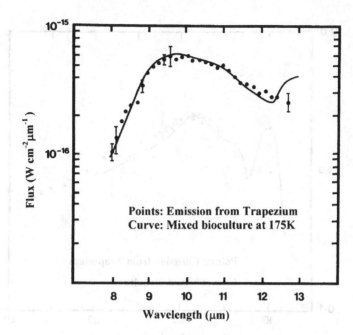

Fig. 11   Observed infrared flux from the Trapezium nebula (points) compared with the emission calculated for a mixed microbial culture including diatoms heated to 175 (curve).

intervals of time, and re-emit radiation over certain infrared lines, including one at 3.28 μm. Needless to say, such molecules are part and parcel of biology, and their occurrence in interstellar space is readily understood as arising from the break-up of bacterial cells.

Fred and I showed much later that galactic infrared emissions at 3.28 μm and other infrared wavelengths, combined with extinction at 2175 Å, can be explained on the basis of an ensemble of biologically generated organic molecules. (F. Hoyle and N.C. Wickramasinghe, *Astrophys. Space Sci.* **154**, 143–147, 1989; N.C. Wickramasinghe, F. Hoyle and T. Al-Jabory, *Astrophys. Space Sci.* **158**, 135–140, 1989.)

As was pointed out in an earlier chapter, even much earlier in 1962, the presence of aromatic molecules in space might have been inferred from the so-called diffuse interstellar absorption bands. It has been known for over half a century that some 20 or more diffuse absorption bands appear in the spectra of stars, the strongest

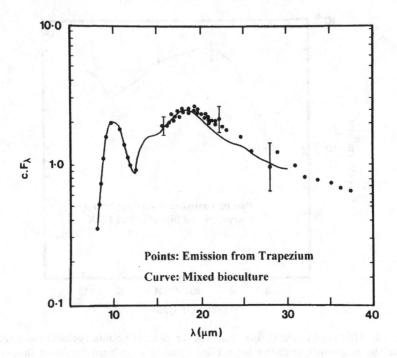

Fig. 12 Same as Fig. 10 but extended to longer infrared wavelengths.

being centred on the wavelength 4430 Å. Despite a sustained effort by scientists over many years no satisfactory inorganic explanation for these bands has emerged. I came across a possible solution at the conference in Troy New York to which I have already referred. The Chemist F.M. Johnson showed that a molecule related to chlorophyll — magnesium tetrabenzo porphyrin — has all the required spectral properties. Chlorophyll of course is an all important component of terrestrial biology — it is the green colouring substance of plants, the molecule responsible for photosynthesis, the process that lies at the very base of our entire ecosystem on the Earth.

Very recently we have unearthed yet another property of biological pigments such as chlorophylls, a property that clearly shows up in astronomy. Many biological pigments are known to fluoresce, in the fashion of pigments in glowworms. They absorb blue and ultraviolet radiation and fluoresce over a characteristic band in the red part of the spectrum. For some years astronomers have been detecting a broad emission feature of interstellar dust over the waveband 6000–7500 Å. Chloroplasts containing chlorophyll when they are cooled to temperatures appropriate to interstellar space fluoresce precisely over the same waveband. (See F. Hoyle and N.C. Wickramasinghe, *Astrophys. Space Sci.* **235**, 343–347, 1996.)

Chapter 14

Life on the Planets

Despite the mounting hostility towards us, we were able to carry on our work whilst maintaining a degree of cheer. This was due to several reasons. There was the fortunate circumstance that the research we were engaged in did not require much in the way of financial support — certainly not remotely like the funding our colleagues used to get from the public purse. Most importantly we had the support of Principal Bill Bevan who from the outset had an instinct that we were on the right track. We could count on him for all our modest financial needs. At a later stage, Bill Bevan introduced us to Gary Weston, Chairman of Associated Foods whose generous support for our work continued well into the 1990's. This latter support was useful, for instance, to provide Fred with his first fax machine to facilitate our communication, and for meeting the escalating costs of our telephone bills. Last but not least, we had the support of our wives. I know that Barbara encouraged Fred to fight in defence of his views as did Priya, who felt we should continue the struggle to win whatever it took!

We were now firmly committed to the view that an immensely powerful cosmic biology came to be overlaid on the Earth from the outside some 4 billion years ago, through the agency of comets. Other planetary bodies, within the solar system and elsewhere, must also be exposed to the same process. Wherever the broad range of the cosmic life system contains a form of life (genotype) that matches a local niche of a recipient planet, that form would succeed in establishing itself. In our view the entire spectrum of life, ranging from the humblest single-celled lifeforms to the higher animals, must be introduced on a planet from the external cosmos. With this in mind we began to examine new data on the planets of our solar system obtained from Pioneer and Voyager spacecraft for telltale signs of microbial life. We took as an indispensable condition for bacterial life the need to have access to liquid water.

Subject to this constraint we discovered what we thought were tentative signatures of bacterial life in the planets Venus, Jupiter and Saturn. Since Venus is exceedingly hot at ground level (about 450°C) it would be impossible for life to exist at the surface. Venus, however, has an extensive cloud cover and it is within these clouds that life may have taken root. Water is present in small quantities and in the higher atmosphere the temperature is low enough for water droplets to form. Moreover the clouds of Venus are in convective motion in the upper atmosphere, which ranges in height between 70 km to 45 km, with a corresponding temperature range of 75°C at the top to −25°C at the bottom. We argued that the survival of bacteria over the range of conditions in the upper atmosphere was possible, and that repeated variations of temperature in a circulating cloud system would tend to favour bacteria capable of forming sturdy spores. We argued for an atmospheric circulation of bacteria on Venus between the dry lower clouds and the wetter upper clouds where replication might take place. We discovered that Pioneer spacecraft data, including the presence of a rainbow in the upper clouds, could be interpreted as implying the presence of scattering particles that had properties appropriate to bacteria and bacterial spores.

These ideas have recently come into vogue. In 2002 Dirk Schulze-Maluch and Louis Irwin looked at data on Venus from the Russian Venera space missions and the US Pioneer Venus and Magellan probes. They discovered trademark signs of microbial life from studies of the chemical composition of Venus's atmosphere 30 miles above the surface. They expected to find high levels of carbon monoxide produced by sunlight but instead found hydrogen sulphide and sulphur dioxide, and carbonyl sulphide, a combination of gases normally not found together unless living organisms produce them. They conclude that microbes could be living in clouds 30 miles up in the Venusian atmosphere, exactly in the manner we discussed 25 years earlier (*New Scientist*, 26 September 2002).

We had also argued for localised bacterial populations in Jupiter's atmosphere that might even have a controlling effect on its meteorology, including the persistence of the Great Red Spot. A kilometre-sized cometary object hitting Jupiter at high speed will be disintegrated into hot gas that would form a diffuse patch similar to the Great Red Spot. Such a region of the atmosphere would be rich in the inorganic nutrients needed for the replication of microorganisms. A large bacterial population could then be built up in this area and the possibility arises for a feedback interaction to be set up between the properties of the local bacterial population and the global meteorology of Jupiter as a whole. Additionally we presented a case for bacterial grains trapped in the rings of Jupiter, Saturn and Uranus, rings that were discovered in 1979 by the Voyager missions.

We had also come to regard the presence of methane and other organic compounds in any quantity on solar system bodies as being an indication of life. Essentially all the organics on the Earth today are either directly or indirectly due to biology. So it is likely to be for organic matter that is found in substantial quantity elsewhere in the Universe. Astronomers have been used to thinking of the presence of methane in the atmospheres of the four large outer planets as being the result of a thermodynamic trend of carbon compounds to change to methane at low temperatures in the presence of a great hydrogen excess. Yet in our model of the formation of the solar system (discussed earlier) there was no great hydrogen excess in the solar nebula: the bulk of the hydrogen and helium in the disc escaped at the periphery carrying away the excess angular momentum that led to the ejection of the disc in the first place.

If one considers a mixture of free hydrogen and carbon dioxide placed in a flask the separate gases will persist unchanged for an eternity. The thermodynamic trend to methane is essentially unobservable. It is in such situations that catalysts come into their own. Of all the catalysts in the Universe, bacteria are by far the most efficient. Methane producing bacteria (methanogens) exist precisely for speeding up the conversion of carbon dioxide and hydrogen to methane and water. If the transformation occurred inorganically there would be no niche for an entire kingdom of bacteria — the methanogens. So the outer planets, which are known to contain methane in their atmospheres, must be teeming with methanogens according to our point of view. All these speculations were discussed at length in a Cardiff Blue Preprint entitled "On the Ubiquity of Bacteria" and were later published in abridged form in *Space Travellers: The Bringers of Life* (University College, Cardiff Press, 1981).

The next fellow traveller to accompany us in our journey was Hans Dieter Pflug from the Geological Institute of Justus Liebig Universtiy in Giessen, Germany. In 1979 Pflug had presented evidence for microbial fossils in the sedimentary rocks of southwest Greenland (the Isua Series). These rocks being dated at 3800 million years put the first appearance of life back by some 500 million years from previous estimates, thereby reducing the time available for the development of any primordial soup. In fact it turns out that before 3800 million years the Earth was subject to severe cometary bombardment, so that Pflug's microfossils could represent the first respite from these impacts and the first opportunity for life on Earth to survive.

Pflug discovered structures in the shape of fossilised cells occurring as colonies and individually in different stages of budding. His technique using thin sections of the rock seemed to be beyond reproach and free of contamination, but in view of the nature of his findings it was inevitable that controversy would ensue. Objectively, however, his case was water-tight. Spectroscopic studies showed organic molecules confirming the conclusion that these structures did indeed show evidence of early life. In one particular instance there was a cell that appeared to possess a nucleus — a eukaryotic cell resembling a yeast cell. This was of course contrary to the prevailing paradigm of biology which states that cells with nuclei came much later in the process of biological evolution on the Earth. Pflug's paper appeared in *Nature* (H.D. Pflug and H. Jaeschke-Boyer, *Nature* **280**, 483–486, 1975). Predictably several rebuttals were to follow claiming that Pfug's microfossils were not relics of biology but crystallographic artefacts. Similar arguments have continued to the present day.

Pflug contacted us in 1980 offering us information that was even more interesting than terrestrial microfossils. He claimed to find compelling new evidence for bacterial microfossils in carbonaceous meteorites. The historical background to this work is worth recalling before describing Pflug's new finds.

As the name implies, the carbonaceous meteorites contain carbon in concentrations upwards of 2 percent by mass. In a fraction of such meteorites the carbon is known to be present as high molecular weight organic compounds. Although there is still some debate on the matter, it is generally held that at least one class of carbonaceous meteorite is of cometary origin. If one thinks of a comet containing an abundance of frozen microorganisms, repeated perihelion passages close to the sun could lead to the selective boiling off of volatiles, admitting the possibility of sedimentary accumulations of bacteria within a fast shrinking cometary body. We can thus regard carbonaceous chondrites (a type of meteorite) as being relic comets after their volatiles have been stripped.

Microfossils of bacteria in meteorites have been claimed as early as the 1930's, but the very earliest claims were quickly dismissed as being contaminants. The story did not end there, however, and the whole argument was revived in the early 1960's. The actors in the new drama included Harold Urey, who was personally known to Fred, and was one of the greatest geologists of the century. H. Urey together with G. Claus, B. Nagy and D.L. Europe examined the Orguel carbonaceous meteorite, which fell in France in 1864, microscopically as well as spectroscopically. They claimed to find evidence of organic structures that were similar to fossilised microorganisms, algae in particular. The evidence included electron micrograph pictures, which even showed substructure within these so-called "cells". Some of the structures resembled cell walls, cell nuclei, flagella-like structures, as well as constrictions in some elongated objects that suggested a process of cell division. These investigators, like their colleagues before them, became immediately vulnerable to attack by orthodox scientists.

With a powerful attack being launched by the most influential meteoritists of the day, the meteorite fossil claims of the 1960's became quickly silenced. One of the more serious criticisms that were made against these claims was that the meteorite structures included some clearly recognisable terrestrial contaminants such as rag-weed pollen. But the vast majority of structures ("organized elements") that were catalogued and described were not contaminants. Intimidated by the ferocious attack that was launched against them, Claus reneged under pressure, and Nagy retreated while continuing to hint in his writings that it *might be so*, rather in the style of Galileo's whispered "*E pur si muove*".

An alternative explanation was that these fossil-like structures were mineral grains which have acquired coatings of organic molecules by some non-biological process. The difficulty with this theory, however, is that the highly organized cell-like appearance of these structures would still remain a mystery. Terrestrial contaminations were a possibility, but this was also unlikely to be the correct explanation for most of the structures, because they have no modern terrestrial counterpart.

In 1980 Pflug reopened the whole issue of microbial fossils in carbonaceous meteorites. Pflug used techniques that were distinctly superior to those of Claus and his colleagues and found a profusion of cell-like structures comprised of organic matter in thin sections prepared from a sample of the Murchison meteorite which fell in Australia, about a hundred miles north of Melbourne on 28 September 1969. He showed these images to us and both Fred and I were convinced of their biological provenance. Pflug himself was a little nervous to publish the results, fearing for his career and anticipating the kind of reaction that was seen in the 1960's. We convinced him to present his work at the out-of-town meeting of the Royal Astronomical Society, held in 1980 in Cardiff, to which I have already referred.

The method adopted by Pflug was to dissolve-out the bulk of the minerals present in a thin section of the meteorite using hydrofluoric acid, doing so in a way that permits the insoluble carbonaceous residue to settle with its original structures intact. It was then possible to examine the residue in an electron microscope without disturbing the system from outside. The patterns that emerged were stunningly similar to certain types of terrestrial microorganisms. Scores of different morphologies turned up within the residues, many resembling known microbial species. An example is shown in Fig. 13. It would seem that contamination was excluded by virtue of

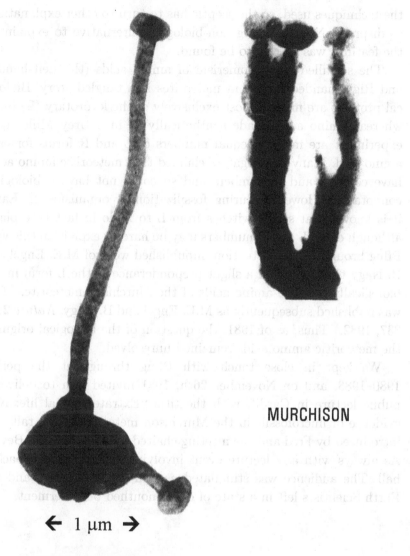

MURCHISON

← 1 μm →

PEDOMICROBIUM
RECENT

Fig. 13 One of Hans Pflug's many microfossils from the Murchison meteorite compared with recent microorganism Pedomicrobium.

the techniques used, so the sceptic has to turn to other explanations as disproof. No convincing non-biological alternative to explain all the features was readily to be found.

The so-called stereoisomerism of amino acids (the Left-handed and Right-handed forms) in meteorites is a tangled story. Biological proteins are made almost exclusively of the levorotary (L) form, whereas amino acids made synthetically as in a Urey–Miller type experiment are made of equal numbers of L and R forms for each amino acid. Many investigators claimed that meteoritic amino acids have equal L and R numbers and so could not have a biological connotation. However, during fossilization of organisms on Earth it is known that some switches from L to R do in fact take place, although equal L and R numbers may be hard to explain in this way. Pflug brought to our attention unpublished work of M.E. Engel and B. Nagy that indicated a slight preponderance of the L form in the biologically relevant amino acids of the Murchison meteorite. (This was published subsequently as M.E. Engel and B. Nagy, *Nature* **296**, 837, 1982.) Thus, as of 1981, the question of the biological origin of the meteoritic amino acids remained unresolved.

We kept in close touch with Pflug throughout the period 1980–1983, and on November 26th, 1981 invited him to deliver a public lecture in Cardiff with the title "Extraterrestrial life: New evidence of microfossils in the Murchison meteorite". The talk was introduced by Fred and the meeting chaired by Principal Bill Bevan. As always with any lecture event involving Fred we had a packed hall. The audience was stimulated as well as entertained, and the Earth Scientists left in a state of open-mouthed bewilderment.

Chapter 15

Evolution from Space

With the amount of evidence we now had for an organic composition of interstellar dust we found it exceedingly puzzling to understand the reluctance to accept even this relatively simple fact. Perhaps there was a perception that very much bigger issues were at stake. If the whole of Darwinian evolution was to come under scrutiny there would be a motive to turn away from even the simplest facts that pointed in such a direction. After all the victory of Darwinism over the Judeo–Christian view of creation as exemplified in the Huxley–Wilberforce debate was a hard-won affair and the memory of the blood-letting must still linger in our collective consciousness. It is a victory to be cherished at all cost, and smaller truths may need to be sacrificed in the interests of larger perceived goals.

On Fred's visits to us we would often touch upon such matters. At times we were philosophical, at other times irate in dealing with the attitudes of our scientific colleagues. One afternoon, when we had tired of figuring out some astronomical problem, we decided to take a stroll to Caerphilly mountain, a few miles from where I lived. We had reached the top of the mountain when a dog darted across the footpath and snapped at Fred, sunk a tooth into his ankle and tore the bottom of a trouser leg before an embarrassed owner came to the rescue. Later that evening the owner came to our house to further apologise and offer compensation and remarked that this was the very first time his dog had behaved in this aggressive way. Fred surmised that his anger at the intransigence of our critics may be the cause, the dog sensing the adrenaline that he produced!

The next major project we undertook during the period 1980–1981 was an attempt to connect cosmic life, viruses and bacteria causing disease, with the evolution of life on the Earth. If life started on the Earth some 4 billion years ago with a comet bringing the first cosmic microorganisms, how did it evolve and diversify to produce the magnificent range of life forms we see today?

It is believed by neo-Darwinists that the full spectrum of life is the result of a primitive living system being sequentially copied billions upon billions of times. According to their theory, the accumulation of copying errors, sorted out by the processes of natural selection, the survival of the fittest, could account for both the rich variety of life and the steady upward progression of complexity and sophistication from a bacterium to man. This is perhaps a simple representation of neo-Darwinism, but it encapsulates its essential features. Is this enough to explain the available facts of biology?

When we began to examine the matter our answer turned out to be an emphatic no. Our mathematical objections to the theory were published in a booklet entitled *Why Neo-Darwinism Does Not Work* (University College Cardiff Press, 1982), and a more general critique of neo-Darwinism in our book *Evolution from Space* (Dent, 1981).

In essence our basic argument was simple. Major evolutionary developments in biology require the generation of new high-grade information, and such information cannot arise from the closed box evolutionary arguments that are currently in vogue. The same difficulty that exists for the origin of life from its organic building blocks applies for every set of new genes needed for further evolutionary developments. One of the earliest arguments in this context concerned the origin of the set of enzymes needed for a primitive bacterium.

A typical enzyme is a chain with about 300 links, each link being an amino acid of which there are 20 different types used in biology. Detailed work on a number of particular enzymes has shown that about a third of the links must have an explicit amino acid from the 20 possibilities, while the remaining 200 links can have any amino acid taken from a subset of about four possibilities from the bag of 20.

This means that with a supply of all the amino acids supposedly given, the probability of a random linking of 300 of them yielding a particular enzyme can be calculated to be as little as 10^{-250}. The bacteria present on the Earth in its early days required about 2000 such enzymes, and the chance that a random shuffling of already-available amino acids happens to combine so as to yield all the required 2000 enzymes is about one in $10^{500,000}$.

Everybody must surely agree that a probability as small as this cannot be contemplated. So to a believer in the paradigm of the origin of a bacterium or indeed a set of genes in the warm little pond, there has to be a mistake in this argument. Although it is known that the bacteria present on the Earth, almost from the beginning, were ordinary bacteria, modern bacteria as one might say, it has been argued by some that the first organisms managed to be viable with considerably fewer than 2000 enzymes. A number of about 256 has been quoted and in this case the probability of origin of this severely sawn-down enzyme set is one in 10^{6900}, still not a bet one would advise a friend to take. For comparison, there are about 10^{79} atoms in the whole visible universe, in all the galaxies visible in the largest telescopes.

Such statistics convinced us beyond any shadow of doubt that life must be a truly cosmic phenomenon. The first origin of the magnificent edifice we recognise as life could not have begun in a warm little pond here on Earth, nor indeed in any single diminutive location in the cosmos. It must have required all the resources of the stars in a large part of the universe to originate, and thence its spread is easily achieved. Many attempts to convey this were made by Fred and myself in our writings and lectures by means of comparison with everyday situations. One such comparison was that if a population of several thousand people were to throw a pair of unbiased dice, the probability of everybody throwing two sixes is of the same order of difficulty as the origin of the set of enzymes needed for a primitive bacterium. Another comparison is that the origin of life from organic molecules on Earth is of a smaller order of improbability than a tornado blowing through a junk yard assembling a fully working Boeing 707!

The many attempts that have been made thus far to overcome this hurdle were in our view wholly unsuccessful. In a lecture delivered by Fred at the Royal Institution on 12 January 1982 he said thus:

> "*Some people are putting out statements that by appealing to a mysterious process called non-equilibrium thermodynamics the problem of finding the required explicit orderings of amino acids can somehow be solved. This is like saying to a person trying to throw a sequence of 5 million sixes that they would do better if there were to roll the dice a bit faster, whereas of course it would scarcely help at all even if the dice were thrown at the speed of light.*"

To overcome these seemingly insurmountable obstacles for the origin of life there seemed to us to be two logical options:

Option 1: The alternative to the assembly of life by random, mindless processes in a finite universe is assembly through the intervention of some form of cosmic intelligence. Such a concept would be rejected outright by many scientists, although there is no purely logical reason for such a rejection. With our present technical knowledge human biochemists and geneticists could now perform what even 10 years ago would have been considered impossible feats of genetic manipulation. We could for instance splice bits of genes from one species into another, and even work out the possible outcomes of such splicings. It would not be too great a measure of extrapolation, or too great a license of imagination, to say that a cosmic intelligence that emerged naturally in the Universe may have designed and worked out all the logical consequences of our living system. If the Universe was of the standard Big Bang type with an age of no more than 13.75 billion years one might well be stuck with such an explanation, unless life itself can somehow beat the improbability factors we have just discussed. For instance, one might assert that the information content of life is deeply buried in the structure of matter at a yet undiscovered subatomic level.

Option 2: The first premise for the origin of life and all its genetic facets is the existence of a spatially infinite universe, a universe that ranges far beyond the largest telescopes. Then the very small chance of obtaining a replicative primitive cell will bear fruit somewhere and, when it does, exponential replication will cause an enormous number of the first cells to be produced. It is here that the immense replicative power of biology shows to great advantage. It generates enough copies of itself for a second highly improbable evolutionary event to occur in one of the profusion of offspring from the first cell. And so by an extension of the argument to the third improbable event. Indeed to a whole chain of improbable occurrences, which result at last in an enormous diversity of cells we have today, the cells that were already present at the formation of the Earth. This is option that Fred himself considered to be the most reasonable, and one that connected with his cosmological preferences.

On this view of the origin of life there would be little variation in the forms to which the process gives rise, at least so far as basic genes are concerned, over the whole of our galaxy. Or indeed over all nearby galaxies. The rest of the story concerns the many ways in which the same basic genes can combine to produce rich varieties of living forms from one environment to another, always remembering that because of the large numbers involved — large numbers of stars, large numbers of planets and large numbers of galaxies, the system can afford many failures. For instance the Earth, in its four and a half billion year history, would not have produced anything very noteworthy but for the chance events of the last half-billion years.

Our views on cosmic evolution connect also with the idea of disease-causing viruses coming from space. The question was asked by our critics: "How could a virus or bacterium coming from space relate to evolved life forms on the Earth?" The answer must be that higher organisms evolved in response to exposure to space borne viruses. Apart from causing diseases, viruses could on occasion add on to our genes, and so provide a store of future evolutionary potential. This would be the *raison d'être* for the persistence of viral diseases in evolved life forms such as ourselves. One might legitimately ask: if virus infections are bad for us why did the evolution of higher life not develop a strategy for excluding their ingress into our cells? Logically it seems easy enough for the greater information content of our cells to devise a way of blocking the effects of the much smaller information carried by a virus, and yet this has not happened in the long course of evolution. Could it be, we wondered, whether this "invitation" to viruses was retained for the explicit purpose of future evolution?

Chapter 16

Theories of Trial

Fred was far from being a religious man, not in the conventional sense at least. His position, as far as I could assess, was that *if* there were a cosmic creator it would be scarcely conceivable that any of the world's religions would have fully grasped either *His* intent or *His* plan. A degree of incompleteness in comprehending such matters must necessarily remain. I believe he kept an open mind, as I did, and regarded "creation", in some form, as being a valid intellectual position to hold in relation to the origin of life. He was also cynical of the ambivalent scientific attitude that prevailed in relation to this whole question: whilst it was considered untenable for "creation" to be used in connection with life, it was perfectly acceptable to contemplate that an entire Universe, with all its inherent laws, suddenly came into existence some 13.75 billion years ago, created to all intents and purposes. If Fred and I ever discussed "creation" or a "creator" we did so only as an abstract concept free of any specific Judeo–Christian implications.

In 1981 we had already published our book *Evolution from Space* which was receiving a great deal of media attention, particularly a chapter with the enigmatic heading "Convergence to God?" It was not surprising, therefore, that Fred was approached by the State Attorney of the State of Arkansas asking if he would give expert evidence for the State at a forthcoming Creation trial. On March 19, 1981, the Governor of Arkansas had signed into law an Act which stated: "Public schools within this State shall give balanced treatment to creation-science and to evolution-science." The US Federal

Government had challenged the constitutional validity of the Act, and the case being heard was the State of Arkansas versus the Federal Government. Fred was not able to oblige due to other commitments, and directed the request to me for consideration. I spoke at length to the Attorney who conyinced me that all I was required to do was to defend the ideas we published in *Evolution from Space*. To be their expert witness I had to rebut the claim of the Federal Government that neo-Darwinian evolution was a proven fact. Although I was a little apprehensive of what I might be letting myself in for, I did not see an immediate reason for declining their invitation. After several long telephone conversations with Fred it was agreed that I should go to Arkansas and present a testimony that we would agree upon beforehand. I had religious friends and I respected peoples' freedom to hold such beliefs. I did not feel that their legitimate aspirations should be thwarted on scientific grounds that seemed to me to be insecure.

My wife Priya, my youngest daughter and I set out to Arkansas on a cold December day in 1981 through a snowbound airport at Heathrow. There were long delays due to snow, and I remember thinking many times that this was an ill omen and that we should turn back and return home. But we made the trip and eventually reached Arkansas just in time for the deposition and the trial. The case I presented essentially summarised my scientific beliefs. The following quotations are an extract of my testimony.

"The facts as we have them show clearly that life on Earth is derived from what appears to be an all pervasive galaxy-wide living system... Life was derived from and continues to be driven by sources outside the Earth, in direct contradiction to the Darwinian theory that everybody is supposed to believe...

It is stated according to the theory that the accumulation of copying errors, sorted out by the process of natural selection, the survival of the fittest, could account both for the rich diversity of life and for the steady upward progression from bacterium to Man... We agree that successive copying would accumulate errors, but such errors on the average would lead to a steady degradation of information... This conventional wisdom, as it is called, is similar to the proposition that the first page of Genesis copied billion upon billions of time would eventually accumulate enough copying errors and hence enough variety to produce not merely the entire Bible but all the holdings of all the major libraries of the world... The processes of mutation and natural selection can only produce very minor effects in life as a kind of fine tuning of the whole evolutionary process...

In our view every crucial new inheritable property that appears in the course of the evolution of species must have an external cosmic origin... We cannot accept that the genes for producing great works of art or literature or music, or developing skills in higher mathematics emerged from chance mutations... If the Earth were sealed off from all sources of external genes: bugs could replicate till doomsday, but they would still only be bugs: and monkey colonies would also reproduce but only to produce more monkeys. The Earth would be a dull place indeed...

The notion of a creator placed outside the Universe poses logical difficulties, and is not one to which I can easily subscribe. My own philosophical preference is for an essentially eternal, boundless Universe, wherein a creator of life may somehow emerge in a natural way. My colleague, Sir Fred Hoyle, has also expressed a similar preference. In the present state of our knowledge about life and about the Universe, an emphatic denial of some form of creation as an explanation for the origin of life implies a blindness to fact and an arrogance that cannot be condoned."

My own testimony which was consistent with my beliefs, and which Fred wholeheartedly endorsed, is not a source of regret in itself. The case was won against the State of Arkansas Education Board whom I was supposed to be representing. In his summing up of the judgement on 5 January 1982, Judge William R. Overton made the following statement:

> "*In efforts to establish 'evidence' in support of creation science, the defendants (The State of Arkansas) relied upon the same false premise..., i.e., all evidence which criticized evolutionary theory was proof in support of creation science... While the statistical figures may be impressive evidence against the theory of chance chemical combinations as an explanation of origins, it requires a leap of faith to interpret those figures so as to support a complex doctrine which includes a sudden creation from nothing, a worldwide flood, separate ancestry of man and apes, and a young earth...*
>
> "*The defendants' argument would be more persuasive if, in fact, there were only two theories or ideas about the origins of life and the world... Dr. Wickramasinghe testified at length in support of a theory that life on earth was 'seeded' by comets which delivered genetic material and perhaps organisms to the earth's surface from interstellar dust far outside the solar system... While Wickramasinghe's theory about the origins of life on earth has not received general acceptance within the scientific community, he has, at least, used scientific methodology to produce a theory of origins which meets the essential characteristics of science.*
>
> *The Court is at a loss to understand why Dr. Wickramasinghe was called in behalf of the defendants. Perhaps it was because he was generally critical of the theory of evolution and the scientific community, a tactic consistent with the strategy of the defense. Unfortunately for the defense, Dr. Wickramasinghe demonstrated that the simplistic approach of the two model analysis of the origins of life is false. Furthermore, he corroborated the plaintiffs' witnesses by concluding that 'no rational scientist' would believe the earth's geology could be explained by reference to a worldwide flood or that the earth was less than one million years old.*"

The repercussions of my court appearance unfortunately lasted for several years. Although I had not compromised my beliefs (when cross-examined, I often had to agree with the plaintiffs' claims), many scientists were angry at what they wrongly perceived as our attempt to give credibility to "creation science" which had come to be regarded as the antithesis to science. It was only after meeting "creation scientists" in Arkansas who believed in the literal truth of the Bible, including a belief in an Earth no older than 6000 years, that I began to doubt the wisdom of our decision. For many years following the trial, my family and I were plagued by death threats from several unknown extremist groups, and together with Fred I bore a heavy burden of ostracism for expressing our views in the Arkansas trial.

It was a great relief at this time to be able to escape from such troubles. From the beginning of 1980's I became involved in academic and scientific affairs in Sri Lanka in my capacity as advisor to President J.R. Jayawardene. In this connection I had to make frequent brief visits to the island, which were quite welcome. President J.R. Jayawardene was my father's contemporary at school and had sought me out through the extensive publicity I was receiving at the time for my collaboration with Fred. He had visited India and had been much impressed by the standards of scientific research that prevailed there. As newly appointed President he set himself the task of revitalising Sri Lankan science. He invited me in the summer of 1980 to work on a blueprint for an Institute of Fundamental Studies, based roughly on the model of the Tata Institute in India, with the difference that the President himself was to be the Chairman of its Board of Management. By January 1983 the Institute was set up and I was invited by "JR" (as he was called) to be its founding Director, whilst still retaining my position in Cardiff.

I went to Sri Lanka in the summer of that year with Priya and our three children and prepared for our longest stay in the country since our marriage in 1966. We rented an apartment in Colombo and made every effort to reintegrate with the social and academic scene. I had secured a substantial grant from the United Nations Development Programme (UNDP) to put the fledgling Institute on the world scientific map by organising an interdisciplinary international conference under its aegis. The aim of the conference was to introduce prominent scientists from around the world to the Sri Lankan scene in the hope that they would be able to forge links with researchers working on the island.

Once I had successfully got together the basic infrastructure of the Institute — rented a building with offices located just outside the heart of the city, hired a secretary, an accountant and other office staffs — I started work on planning the conference. I proceeded to do this by involving local academics in Sri Lanka whilst also involving Fred closely on decisions as to who we should invite. How much more interdisciplinary could one get than the areas of research we ourselves were currently engaged in? We thus found a rationale for including an extended session relating to our immediate research interests. Our list of invitees included Zdnek Kopal, Gustav Arrhenius (grandson of Svante Arrhenius), Hans Pflug, Bart Nagy (who was involved in the controversy over organised elements in meteorites), Keith Bigg (an atmospheric physicist who had found bacteria-like structures in the stratosphere), Sri Lankan expatriate scientists Cyril Ponnamperuma and Asoka Mendis, Phil Solomon, Arthur C. Clarke, Tom Gehrels, Jayant Narlikar and Arnold Wolfendale (then President of the Royal Astronomical Society).

The participants began to arrive a few days before the conference in December 1982 and were all accommodated at the Lanka Oberoi, a five star hotel in Colombo. The meeting itself was to take place at the Bandaranaike Memorial International Conference Hall (BMICH), a well-equipped auditorium and conference facility that had been gifted to Sri Lanka by the People's Republic of China. Fred arrived in Colombo from Sydney after an extended visit to Australia. The previous summer we had prepared a preprint entitled "Proofs that Life is Cosmic" which contained what we considered to be the arguments from many different disciplines, all pointing to the cosmic origins of life. We had decided to publish this document as a *Memoir of the Institute of Fundamental Studies, Sri Lanka, No.1*. Our manuscript was delivered to the Government Printer of Sri Lanka to be printed as a government publication on the instructions of the President himself. Within days, Fred and I found ourselves holed up together in the Printer's office — a dingy, Dickensian space — to correct the proofs. We were pleasantly surprised to find very few typos and the book soon saw the light of day.

The conference itself was a high profile national event in Sri Lanka with President Jayawardene delivering an opening address, followed by a keynote lecture given by Fred Hoyle. Fred gave as usual a brilliant exposition of our theories in the unusual circumstance in which a Head of State was in the audience.

One might have thought that the conjunction of talks by Pflug on the Murchison microfossils, Bigg on microbes in the upper atmosphere, Nagy on D/L ratios of amino acids in meteorites in the presence of Gustav Arrhenius (grandson of Arrhenius), may have struck a chord of consonance. But this was not to be. Hans Pflug cautiously presented his slides as he had done on several occasions in the past, and stated the barest of facts without making any inferences. Likewise Bigg presented his intriguing pictures of particles resembling bacteria in the atmosphere, shyly and with minimal commentary. No sooner than these presentations were completed, Gustav Arrhenius then set upon both them with vengeance, claiming that all such finds had to be interpreted as non-biological artefacts. It seemed that he was determined to turn his back on grandfather Svante's old ideas of panspermia that he considered to be improper. There was clearly no way of winning him over, no matter how strong the arguments in our favour might turn out to be. People do indeed see what they want to see and don't see what they don't.

Bartholomew Nagy's case was different. He knew that he had discovered something profoundly important, first with the microfossils and later with the D/L ratios of amino acids in meteorites. In his formal presentation in Colombo his delivery was strident and straightforward, but when Fred and I met him in the hotel bar and engaged him in conversation it was evident that he was an exceedingly frightened man. As Fred put it *"like a rabbit who was being hunted"*.

The sessions that dealt with the question of cosmic origins of life were concluded without any resolutions of the main issues that were raised. Arnold Wolfendale (then President of the Royal Astronomical Society) who was a silent observer of these sessions assured us that he would arrange a discussion meeting at the RAS to continue the debate in London.

Apart from the arguments and conflicts over the Life issue, the conference was enjoyed by everyone. The traditionally lavish Sri Lankan hospitality seemed to surpass itself and there was plenty of time for socialising and relaxing. On the final weekend, a fleet of black Mercedes cars, provided by the Foreign Ministry, arrived at the hotel to take participants on a 3-day cultural tour of the island.

After the formal part of the conference was over, President Jayawardene asked me if I would bring Fred over to President's House, first for lunch and later for a discussion at his offices. The question raised at our meeting was what should be done with the new Institute after the inaugural conference was over. It had become clear at this stage that I would not be able to make a long-term commitment as Director of the IFS in view of the research work I continued to be engaged in. Fred suggested the name of C.J. Eliezer, his old Cambridge friend and student of Dirac, who was at the time a Professor in an Australian University. Jayawardene was visibly disturbed by this suggestion but Fred could not understand why. I had later to explain to him that Eliezer was a Tamil, and Jayawardene was fearful of any links that he might have with the so-called Tamil Tiger movement (LTTE) that was threatening to destroy the integrity of the island. All Tamil intellectuals working abroad were regrettably on a suspect list as far the Government was concerned.

One of the highlights of Fred's brief stay in Sri Lanka was a visit to meet the legendary science fiction writer Arthur C. Clarke. Arthur Clarke had adopted Sri Lanka as his home since 1956 to pursue a passion for deep sea diving and also to write his long series of books including *2001: A Space Odyssey* in delightful surroundings and reflective solitude. Besides his prolific output in science fiction Clarke is also known as the inventor of the idea of the telecommunication satellite. Way back in October 1945 he wrote an article in *Wireless World* that was to change the world forever. He pointed out that a network of satellites in geostationary orbit (Clarke orbit) at a distance of 35786 km above the Earth's surface could serve to give worldwide telecommunication coverage, overcoming the problem of the curvature of the Earth faced by the old telecommunication masts. His ingenious idea is now widely exploited throughout the telecommunication industry, and thanks to Arthur we can make instant contact in sound and pictures across the world.

I first met Arthur in 1962 in an airplane on a journey from London to Colombo and have kept in touch with him ever since. Whenever I go to Sri Lanka I always visit him to talk about astronomy and space and to exchange news. He is a witty and scintillating conversationalist who is never shy of voicing an iconoclastic point of view, and I am always entertained in his company. Arthur was always strongly supportive of my work with Fred, and as our ideas evolved towards cosmic life his support grew stronger. He always had an instinct that these ideas *must* be right, and this was a great source of encouragement to me. Fred and Arthur also shared a publisher and had many common interests, both in science and science fiction. So it seemed appropriate that they should meet.

Whilst we were chatting in Arthur's air-conditioned study in Barnes Place, Fred's eyes alighted on a copy of *Diseases from Space* that was on a shelf in the library. Arthur then made an extremely beguiling comment. He said that he had recently been visited by someone high up in the CIA who had remarked that "they" had evidence to support our view that bacteria come from space. Since then we discovered that in the 1960's, NASA had supported a series of balloon flights into the stratosphere to heights above 40 km and had recovered viable microorganisms that could be cultured by relatively simple means. Their results showed that there were 0.01–0.1 viable bacteria per cubic metre of air, and that the density appeared to increase with height. Because it is exceedingly unlikely that bacteria could be lofted to such heights and in such large numbers, the conclusion that they came from space must surely have drifted even momentarily into the heads of the experimenters — an idea totally alien to the belief system that prevailed. Of course in the 1960's the conditions under which such experiments were conducted could have left room for the objection of possible contamination. But the findings may have worried NASA nevertheless and they dealt with the situation expeditiously by withdrawing support for further flights.

We shall return to similar experiments carried out by ISRO in the year 2001 in the last chapter of this book. But in the meantime Arthur Clarke's personal stand in the "Life from Space" debate was made quite clear to us in Colombo in December 1983. He was, of course, firmly on our side.

When all our visitors had left I spent a few more months in Sri Lanka trying to find ways of making the Institute of Fundamental Studies a long-term success. With local jealousies, amongst other factors, this goal was turning out to be more difficult than I had expected. Our time in Sri Lanka effectively ended in July 1983 when the most savage communal riots in the country's history unexpectedly broke out. I was in audience with President Jayawardene at the time when news broke of looting and arson in the south of Colombo, and our meeting was abruptly terminated. The riots evidently began as retaliation for a Tamil Tiger ambush of an army patrol in the north of the country leaving 13 soldiers dead. The Sinhalese retaliated violently and riots continued throughout the island for several weeks.

Ever since this time the Tamil Tigers, who are seeking to establish a separate Tamil state in the north, have carried out sporadic acts of violence and suicide bombings mainly directed upon Government targets. They were responsible for the murders of several prominent politicians, including Rajiv Gandhi, the Indian Premier, President Premadasa (who succeeded J.R. Jayawardene as President) and several ministers of state in the Government of Sri Lanka. This was not a country which one could have felt comfortable to work in [In 2009 the Tamil Tigers were militarily defeated by the Sri Lankan Government led by President Mahinda Rajapakse. At long last peace and tranquility has returned to the island — a scene totally different from what I saw in 1983].

By 1983 Cardiff Astronomy was advancing in diverse ways. There was a noteworthy development under Bernard Schutz on Relativistic Astrophysics that led eventually to a major group devoted to the search for gravitational waves — a prediction of Einstein's theory of relativity. In our own particular areas of research, which were by now getting increasingly distant from the rest, we had several students, mostly from Iraq, working alongside Shirwan Al-Mufti on various aspects of the biological grain thesis. Niama Jabir was doing more detailed work on theoretical modelling of the interstellar extinction curve (N.L. Jabir, F. Hoyle and N.C. Wickramasinghe, *Astrophys. Space Sci.* **91**, 327–344, 1983) and Laith Karim was studying spectra recorded by the International Ultraviolet Explorer Spacecraft (1981–1982). Karim was accessing this data from the Rutherford Appleton Laboratory to search for details in the extinction curve around 2800 Å for stars which were known to have relatively weak 2175 Å absorption features. This was of interest because DNA or RNA, if it existed in any quantity in free form or as viruses or viroids, would exhibit an absorption band centred at 2600–2800 Å. We did not have great expectations of finding such a band because it is a relatively very weak band and would tend to be submerged in the wings of the 2175 Å feature. Moreover, the effect had not shown up in bacteria studied by Al-Mufti in the laboratory, spectra that were found to have an absorption that peaked at about 2200 Å (F. Hoyle, N.C. Wickramasinghe and S. Al-Mufti, *Astrophys. Space Sci.* **111**, 65–78, 1985). Karim, however, was able to find IUE spectra where the 2200 Å was weak, in which with an appropriate comparison star, a hint of a 2800 Å absorption feature was seen. Looking over his work at the time we did not see any obvious problems with the analysis, so both Fred and I permitted the appearance under joint authorship of a paper reporting this provisional discovery (L.M. Karim, F. Hoyle and N.C. Wickramasinghe, *Astrophys. Space Sci.* **94**, 223–229, 1983).

At the present stage of our progress towards panspermia any technical error we might commit could all to easily become a hostage to fortune. Unsubstantiated opinions, on the other hand, were easier to cope with. It was only after Karim's work was published that we realised that there might be a problem with the data that was being used. An examination of the IUE documentation belatedly revealed that the spectra over the wavelength range of interest suffered from a problem known as "saturation". Under such circumstances no conclusion about the spectra was possible. This unfortunate glitch meant that we had to disregard any effect such as a 2800 Å absorption band in our stars, although this did not pose any threat to our overall model of the grains. But it was an error that our adversaries could seize upon.

After the heat of Colombo conference, the next testing ground for our ideas was a discussion meeting of the Royal Astronomical Society which took place on 11 November 1983. As promised to us in Colombo, the meeting was initiated by Arnold Wolfendale. The discussion had the title: "Are interstellar grains bacteria?" Apart from myself and Fred there was Hans Pflug, Max Wallis and Phil Solomon on one side of the debate, and Mayo Greenberg, Harry Kroto, and Doug Whittet on the other. After Fred and I had presented our evidence in support of an affirmative answer to the question at issue, the others set out to argue the case against. In our view, the case against and the rebuttals were largely polemical.

Quite predictably Mayo Greenberg pounced on the work of Karim that was mentioned earlier, implying that if we got one thing wrong everything that we said had to be dismissed. Doug Whittet raised several objections to the bacterial grain model based on availability of constituent atoms in interstellar space. The various atomic species — carbon, oxygen, nitrogen etc. — present in the interstellar gas and in the dust must together add up to what we know to be overall cosmic abundances. Those elements that go to make up the dust must therefore be depleted from the gas phase. Whittet's first argument was that recent estimates of carbon and oxygen depletions from the interstellar gas were inadequate to allow for the carbon and oxygen that were needed to be tied up as bacterial grains. This argument was shown by Phil Solomon, who was present, to be insecure and probably wrong. More recent studies have shown that Phil was correct and the measured carbon and oxygen depletions are indeed consistent with the bacterial model of dust. Whittet's next point was that the interstellar abundance of phosphorus (present in DNA) was inadequate to support the bacterial model — in other words there was not enough phosphorus around. This was also shown by us to be wrong. If one takes solar abundances as being strictly correct for interstellar gas, and if about 2/3 of a percent of the dry weight of bacteria is taken to be DNA we could face a phosphorus deficit by a factor of between 5 and 10. But not all the interstellar bacteria can be assumed to be viable, and nutrient-starved bacteria are known to have phosphorus deficits.

David Williams provided new data on hydrogenated films of amorphous carbon and argued that this material too could be used to explain the 2.9–3.5 µm spectrum of the galactic centre source GC-IRS7. On closer inspection we did not think the fits were good enough to compete with the bacterial model. Finally, Harry Kroto (now Professor Sir Harry Kroto, Nobel Laureate for discoveries connected with C_{60}) presented the standard chemist's point of view. Nothing that is definite about precise chemical compositions can be inferred from IR spectra, it was argued. Although our fit of a bacterial model to the spectrum of interstellar dust cannot be denied, a chemist's contention is that a combination of organic absorbing groups could have arisen inorganically in just the right proportions to mimic a bacterial spectrum. That was not an argument we heard for the first time, and our answer has been already documented in this book.

With the exceedingly wide range of spectroscopic data now available that relates to interstellar dust, any synthetic mixture of chemical groups that mimics biology and its degradation products cannot be easily envisaged. Biology has the unique property of reproducing with unerring accuracy an entire suite of biochemical substances and structures in precisely the proportions that are demanded by astronomy. To invoke non-biological process to produce the same outcome, and on a vast cosmic scale is, in my view, a desperate attempt to maintain an Earth-centred perspective on life.

Chapter 17

A Fossil Controversy

It would be difficult to continue my story without reference to a strange development in 1983 when Fred was cheated of a Nobel Prize. There is little doubt, even in the minds of his arch enemies, that his work on the origin of the elements in the 1940's and 1950's constitutes a monumental contribution to science. I have mentioned earlier that theory of nucleogenesis (the synthesis of elements in the hot interiors of stars) was an outstanding scientific landmark of the 1950's, and Fred's role as leader and pioneer of this entire venture is beyond question. Fred's early calculations had shown that in order for carbon to be produced in adequate quantities in stars the nucleus of the carbon atom had to possess an excited state, and precisely this level was later discovered in the laboratory by Willy Fowler and a Bob Whaling. In the further development of the theory he collaborated with Willy Fowler and with Geoffrey and Margaret Burbidge in the mid-1950's. In their classic paper B^2FH the four authors published a comprehensive account of stellar nucleosynthesis that remained a cardinal influence in astronomy over many decades.

The 1983 Nobel Prize for Physics was awarded to William A. Fowler and Subramanyam Chandrasekhar for contributions to nucleogenesis. Why Fred Hoyle was excluded in this award remains an inexplicable mystery. It is ironical that some months before the award was announced one of Fred's granddaughters had asked Willy Fowler (who was a family friend of the Hoyles) to write an article for a school magazine. In it Fowler stated that Hoyle was the pioneer of all this work and indeed the main driving force.

There have been, of course, other instances of breaches of justice in regard to acknowledgement of pioneering work on a lesser and a greater scale. The inherent weaknesses of human nature, fraught with jealousies and prejudices, often interferes with objective judgement, and Fred would have been the first to recognise this fact. Fred was also well aware that his own place in the history of science was secure, so that the opinions of politicised academies of science were largely irrelevant in a longer-term perspective of things.

I believe, however, the Nobel incident affected Fred. And although he rarely talked about it, he never talked again to his old friend Willy either. One perceivable effect of the episode was that he was even more vituperative in his attacks on the scientific establishment. There have been many speculations as to the reasons for Fred's exclusion for a prize for his own work. One curious speculation came from John Maddox, Editor of Nature, who surmised that the reason for the exclusion was Fred's involvement in panspermia, a theory that the Swedish Academy did not wish to endorse.

Despite such setbacks, our efforts in the next few years were focussed towards fine-tuning our theory of panspermia as well as dealing with criticisms that were levelled against it. There was a growing consensus that interstellar grains were of a complex organic nature, and this situation we regarded to be a victory. But the identification of interstellar grains with bacteria was still vigorously challenged in many quarters. As described in the last chapter it is our claimed uniqueness of spectral fit to the galactic infrared source GC-IRS7 that came under the closest scrutiny. Our line of defence has always been that the fit proposed was not just for positions of a few absorption peaks over the relevant infrared waveband, but for the entire opacity function $\tau(\lambda)$ over essentially an infinity of wavelengths. The latter requirement was clearly far more stringent, and it was precisely for this reason that properly calibrated laboratory experiments were specially designed. Whether there were other combinations of functional chemical groups in an abiotic system that did the job equally well was always open to question. If an appropriate mixture of such functional groups can be obtained non-biologically, one is faced with the dilemma of explaining how it can occur with the same relative proportions with infallible accuracy throughout the galaxy.

The challenge of finding such a mixture to explain the spectrum of GC-IRS7 was considered so important that several laboratories in the United States and elsewhere began to devote their energies to this task. Such attempts were at best only partially successful.

Fred took pride in his Yorkshire roots, and as a rather forthright Yorkshireman he tended to be impatient when dealing with his critics, particularly if he felt sure that he was right and they were wrong. His responses to false criticisms were often far more aggressive than they might have been, and this situation may have worsened after the conduct of the Swedish Academy. In one instance he commented on a paper by M.H. Moore and B. Donn (*Astrophys. J.* **257**, L47, 1982) which had claimed that an organic residue extracted after irradiating a mixture of inorganic ice possessed properties which could rival our bacterial model in the 3–4 μm infrared waveband. Their laboratory spectroscopic data were displayed in their paper in a way that made it difficult to verify their claim. In a Cardiff preprint under our joint authorship entitled "From NASA with love" he wrote thus:

> "*When any organisation is first puffed-up with a gross over-supply of funds that are subsequently reduced to a moderate over-supply, the organisation always finds itself consuming its entire resources in overheads. This is true even when the supply remains at hundreds of millions, or billions, of dollars a year. NASA today is reduced to a situation in which its overheads are so demanding that it can no longer afford a straightforward sheet of paper . . . in which the researchers were obliged to publish their spectra so shrunken in scale that you would put any one of them on a decent-sized postage stamp . . . At this stage we could take note of the statement of Moore and Donn: 'In the 3.4 μm region, the spectrum of our laboratory-synthesized residue matches closely that of E. coli' . . . A sad situation indeed, for here we have the once-opulent NASA so reduced in circumstances that it can no*

*longer manage to send its researchers on a desperately-needed
visit to the nearest oculist...*"

In other publications Mayo Greenberg and his colleagues had claimed
that a variant of the processes used by Bert Donn and Moore could
lead to trace quantities of a complex polymer, that he called "yellow
stuff" for a better description. The spectrum of this material was also
claimed as another rival to our bacterial spectrum. The repeatability
of all such experiments to produce exactly the same composition of
end product always worried us, and this we thought to be a short-
coming of the abiotic solutions that were being offered. We were keen
to test it out but this was not possible. Our spectrum of a desiccated
bacterial cell, however, was everywhere reproducible by the simple
process of biological replication. In a comment addressed to Mayo
Greenberg, Fred wrote:

*"If Professor Greenberg would be good enough to provide us
with a milligramme of his yellow stuff, we shall in return offer
him a bucketful of horsedung!"*

The horsedung being replete with *E. coli* of course.

In my journey with Fred we sometimes took diversions that led us
through treacherous paths. With hindsight some of these excursions
may better have been avoided. One such diversion, which extended
over 3 years, was concerned with the famous fossil of *Archaeopteryx*.
When we were thinking about evolution, *Archaeopteryx* appeared a
somewhat discrepant oddity in the fossil record. It has been considered to be a link between reptiles and birds and its relevance to the
neo-Darwinian theory of evolution is obvious. When it was discovered it was hailed as one of the long sought-after missing links in the
fossil record. The fossil, ostensibly of a small reptile with exquisitely
well preserved feather impressions, was discovered in a 160 million
year old limestone deposit from the Jurassic period. The discovery
was made in 1877 by Ernst Habelein, a German doctor, in a quarry
at Solnhofen, some 40 miles south Nurenberg. A few years earlier the
same doctor had provided the palaeontological community with a single feather impression on limestone also of the same age, and from the

same quarry. The limestone slab containing the full *Archaeopteryx* fossil, as well as its counter slab (containing the mirror impression of the fossil) were sold to the British Museum, who have prized it as one of their most precious possessions. Little wonder then that even the slightest attempt to challenge its authenticity would provoke a furore.

In September 1984 Fred received a letter, sent to him c/o the Royal Society, by Lee M. Spetner in Rehovot, Israel. Spetner introduced himself as a friend of the distinguished Israeli physicist Cyril Domb with whom Fred had worked closely during the Second World War (1941–1945) on the development of Naval radar. I believe it was Fred's respect for Cyril Domb that made him take Spetner's letter seriously in the first instance. Spetner wrote:

> *"For several years I have had a strong suspicion that the Archaeopteryx fossil is not genuine . . . I suspect that the fossils were fabricated by starting with a genuine fossil of a flying reptile and altering it to make it appear as if it originally had feathers . . ."*

Fred passed this letter on to me, and we were both naturally intrigued by the suggestion. We promptly headed to our nearest libraries (the internet was not available yet) to learn all we could about the history of this fossil. It appeared that fossil forgery was quite commonplace in those days, so for museums to be sold forgeries was by no means an absolute impossibility. There were at least three *Archaeopteryx* related fossils on record from the same source: the single feather (just referred to), the specimen at the British Museum, and another similar specimen in Germany. If the one fossil turned out to be a forgery, the likelihood was that they all were.

A few weeks later Spetner sent us a detailed manuscript in which he summarised his concerns about the authenticity of *Archaeopteryx*, and we felt at the time that he had a *prima facie* case for his thesis. We met Spetner shortly afterwards when he visited Cardiff with his wife, and both Fred and I formed the opinion that he was an honest man — an orthodox Jew who lived by the Book. In fact he turned out to be so orthodox that when my wife Priya entertained him to dinner she discovered that he did not eat any food that was cooked in a non-Jewish home. We had to provide him with raw carrots and apples, which was evidently all he was permitted to eat.

Without indicating what our motives were, we now approached Dr. A.J. Charig at the British Museum for permission to photograph the fossil. Permission was duly granted. In the afternoon of 18 December 1984 we went over to London with the Physics Department's photographer R.S. Watkins and took hundred of pictures of both the slab and the counterslab under various lighting and exposure conditions. When we studied the pictures we saw many features that convinced us that Spetner's case was one that had to be taken seriously. The perfect feather impressions looked as though they were impressed on a thinly applied overlayer (limestone cement) that looked markedly different in texture from the rest of the fossil. We also found that the slab and the counterslab did not match in certain crucial areas. Playing the role of amateur detectives we pored over our pictures for hours on end, and after several months we were sufficiently convinced to go into print. We published our pictures with accompanying comments raising some doubts about the authenticity of the fossil in several articles written for the *British Journal of Photography*. (R.S. Watkins *et al.*, *BJP*, Volume 132, Issues of March 8, March 29 and April 26, pages 264, 358, 468, 1985). The Editor of the British Journal of Photography, Mr. Crawley, issued press releases highlighting the articles, and as a consequence the whole affair received far more media publicity than we might have desired.

Fred's interest on the many puzzles that seemed to be associated with our December 1984 photographs prompted many telephone calls and a couple of visits to Cardiff. Our studies of the pictures as well as our delving into the history of the fossil occupied several months and led to the publication of perhaps our most controversial book, *Archaeopteryx: The primordial bird — A case of fossil forgery* (Christopher Davies Publishers, Swansea, 1986). Just when I thought the fuss was all over, in the late Spring of 1985 I receive a transatlantic phone call from Fred who was visiting the Museum of Natural History in Washington DC. He had discovered that this Museum had a cast of the main slab of the London Archaeopteryx fossil, but not of the counterslab. This may have been quite innocent, but Fred was inclined to think it was not. He asked me to arrange another photographic session to resolve some issues connected with what he saw. I arranged this for 2.30 pm, 23 May 1985, but when Fred and I arrived at the Museum we found a hostile reception awaiting us. The message was simple: they had had enough and would not permit any further access. In fact the Museum went to the trouble of mounting a public exhibition to set out their case against the forgery claim. Finally they thought a line could be drawn under an ugly interlude in their history when they published a rebuttal in a high-impact scientific journal (A.J.F. Charig *et al, Science* **232**, 622, 1986). But all this was not enough to convince either Fred or Spetner. Eventually Spetner managed to secure a few minute samples of the fossil, one from the winged area, another from outside. He carried out scanning electron microscope studies and chemical analyses in Israel and arrived at the conclusion that there were differences between the "suspect" areas and the rest of the fossil. However, the smallness of the samples that were examined still left room for considerable doubt.

The upshot of all this was that nothing was decisively resolved. We did not convince our opponents as we had set out to do, and we also lost many friends! The prudence of taking on so powerful an institution like the British Museum must on retrospect be called to question. Particularly since the outcome of our intended inquiry, whichever way it went, would have had no bearing whatsoever on the bigger issues at hand. Notwithstanding this setback, our progress continued unerringly in the direction of cosmic life. Don Brownlee had begun his programme of collecting cometary dust particles using high-flying U2 aircraft that swept through the lower stratosphere at a height of 15 km (J.P. Bradley, D.E. Brownlee and P. Fraundorf, *Science* **223**, 56, 1984). The method employed was a "flypaper technique" where a sticky plate swept through large volumes of air and captured aerosols as they struck the surface at high speed. Fragile structures like clumps of bacteria or volatile dust would have unfortunately been destroyed by this procedure. What was recovered were mostly porous siliceous clumps with some embedded organics, but occasionally entire organic structures were found buried within them.

Terrestrial particles were easily separated from particles of cometary
origin by studying isotope ratios. (Several key isotope ratios such as
$^{12}C/^{13}C$ are different in comets and terrestrial material). When we
looked at the published pictures we found at least one clear case of
an embedded bacterium-like organic structure with minute embedded magnetite substructures. This structure also turned out to be
uncannily similar to a well-recognised fossilised iron-oxidising bacterium in the Earth's sediments dated at 2000 million years. Since the
latter was found by Hans Pflug, we got together with him and published this comparison in a paper entitled: "An object within a particle of extraterrestrial origin compared with an object of presumed
terrestrial origin" (F. Hoyle, N.C. Wickramasinghe and H.D. Pflug,
Astrophys. Space Sci. **113**, 209–210, 1985) (Fig. 14). This, in our
opinion, was the first strong indication that cometary particles with
a biological provenance are still entering the Earth's atmosphere.
Very recent studies of similar stratospheric particles by C. Floss and

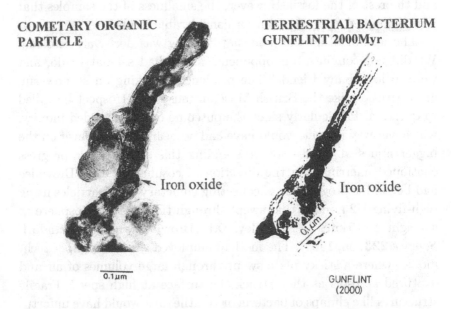

COMETARY ORGANIC PARTICLE

TERRESTRIAL BACTERIUM GUNFLINT 2000Myr

Iron oxide

Iron oxide

0.1μm

GUNFLINT (2000)

Fig. 14 Comparison between an organic particle collected in the lower stratosphere and an iron oxidizing bacterium in 2000 million year old terrestrial sediments. (F. Hoyle, N.C. Wickramasinghe and H.D. Pflug, *Astrophys. Space Science*
113, 209–210, 1985.)

his colleagues (*Science* **303**, 1355–1358, 2004) have shown the presence of heteroaromatic organic compounds with anomalous carbon and nitrogen isotope compositions attributed to a cometary origin. It is ironic that the authors claim that this constitutes evidence for comets seeding the Earth with the complex building blocks of life, exactly the situation we had maintained in 1977.

During 1984 and 1985, both Fred and I, on separate occasions, visited the NASA Marshall Space Flight Centre at the invitation of Richard B. Hoover who later became the Head of Astrobiology at this Centre. In 1985, Hoover's astrobiological interests were embryonic, and I suspect it was his interaction with us that made him move further towards the position he now holds. Richard and his wife Miriam had studied diatoms for many years, and were beginning to become intrigued by some of their bizarre properties. Diatoms are a group of golden-brown algae that have intricately woven silica shells. They are the most dominant microbial life form in terrestrial ice ecosystems such as the Antarctic, and all together they constitute a major component of all marine phytoplankton. Their unearthly properties that the Hoovers pointed out to us include an ability to survive very long periods of desiccation, an ability in some cases to live in total darkness, as well as to endure ionising radiation. In the latter context we learnt that many diatom species are capable of living in environments that contain extremely high concentrations of normally lethal radioisotopes such as americium and strontium. Diatoms thrive in highly radioactive waste ponds including the infamous U-pond. Moreover, they do not merely live here, they actually concentrate radioactive isotopes from an environment! The final surprise was that diatoms appear abruptly in the fossil record 112 million years ago during the late Cretaceous period. This to us was a clear indication that they came from space at this time.

In a joint paper with the Hoovers, Fred and I argued for diatom habitats in ice-water interfaces in comets as well as in the multi-cracked surface domains of the Jovian satellite Europa (Richard B. Hoover *et al.*, *Earth, Moon and Planets* **35**, 19–45, 1986). This paper we believe contains the first attempt to identify the characteristic orange coloration of the cracks of Europa as arising from biological pigments. Later these ideas were taken up by Brad Dalton of NASA's Ames Research Centre. He has argued that visible and infrared spectra of some pigmented extremophilic bacteria can explain the observations of Europa (see *New Scientist*, 11 December 2001). Again we might have been nearly two decades ahead of our time.

Chapter 18

Comet Halley and its Legacy

Fred's visits to Cardiff were always major family events. During his stay with us he would find time to discuss matters that were far removed from science and as my children grew older they too came to appreciate his rich and diverse company. Fred had an unerring interest in classical music and on some mornings (he was an early riser), I would find him in the living room listening intently to Beethoven's Fifth Symphony or Mozart's Death Requiem — music that I recall hearing on so many occasions blaring out from a gramophone at 1 Clarkson Close. He had grown up in the midst of music, his mother being a gifted piano teacher who had studied at the Royal College of Music, and he himself a paid chorister at a local church. Our home in Cardiff too tended to be filled with music as Priya and our three children share a passion for music. Whenever my elder daughter played the piano, Fred would stop whatever he was doing and praise her talents. When at a later date a publisher suggested that she edit our next joint book, Fred readily agreed saying that he would happily entrust such a job to anyone who played the piano so sensitively and so well! The book, however, never materialised.

An amusing incident took place when his birthday (June 24) happened to fall during one of his visits. Priya and I had organised a dinner party to celebrate the event at the "The Walnut Tree", near Abergavenny. Situated at the foot of the Skirrid Mountain and endorsed for its cuisine by Elizabeth David, it was arguably the best restaurant in Wales. We had invited a few other astronomers and friends to join us and the occasion turned out to be most memorable.

At the end of the evening the waiter brought me the bill to pay. I produced my visa card as I always do on such occasions, and to my horror was informed that they accept only cash or cheques. This moment of embarrassment was aggravated by a request to write down my name and address to be shown to the proprietor for appropriate action. Within minutes an ecstatic Franco Taruschino rushes to our table and hugs Priya! Evidently he knew Priya from a recent cookery book she had published. Fred Hoyle and all the other astronomers at our table were unknown quantities as far as he was concerned. His excitement at seeing Priya was so great that he went back in and woke up his young daughter to introduce her. All's well that ends well — after a round of complimentary liquors, we were sent home as friends of VIP Priya, and asked to post a cheque whenever we had the time!

We were now occupied with work for two books, *Living Comets* (F. Hoyle and N.C. Wickramasinghe, University College, Cardiff Press, 1985) and *Viruses from Space* (F. Hoyle, C. Wickramasinghe and J. Watkins, University College, Cardiff Press, 1986). For *Living Comets* we considered all aspects of comets that seemed to have a bearing on life, in particular the question of radioactive heating of comet cores that would permit bacterial replication to occur in the early history of the solar system.

In *Viruses from Space*, we proceeded to update our earlier arguments and also collaborated with a General Practitioner Dr. John Watkins who kindly provided us with data from his family practice in Newport. Looking through his case notes dating from 1970, Watkins identified 16 pairs of twins with ages ranging from six months to 14 years to determine how they succumbed to influenza during epidemics. Of the 118 instances when one twin was diagnosed with acute upper respiratory tract infection (presumed influenza during epidemics) the other twin was found to succumb only in 28 instances. The implied cross-infection rate was only 24%, very close to the attack rate that prevailed in the populace at large. A transmission probability as low as 0.24 would, at any rate, be quite insufficient to explain the facts relating to most influenza epidemics. In another project Watkins confirmed an earlier result that had been obtained by a Cirencester GP, Dr. Edgar Hope-Simpson, from his family practice data. During 1968/69 and 1969/70 Hope-Simpson considered a set of families where at least one member reported with influenza-type illness. In this group he looked at the incidence of subsequent cases on days 1, 2, 3, 4 etc. after the index case. From this data he found that the probability of a second member succumbing to influenza was no more than the attack rate in the community at large. Thus being a member of an "infected" household did not appear to increase the risk significantly. All this confirmed to us that in order to explain the pattern of influenza epidemics some trigger — perhaps a biochemical trigger — if not the entire virus must fall through a turbulent atmosphere and reach the ground in an exceedingly patchy distribution. Whether one succumbed or not during an epidemic depended mainly upon one's location in relation to a general infall pattern.

The next high point in our journey was connected with the return to perihelion of Halley's comet in 1986. This was the first time that a comet was being studied by scientists since the beginning of the space age. From as early as 1982 a programme of international cooperation to investigate this comet came into full swing, the objective being to coordinate ground-based observations, satellite-based studies, and space probe analysis on a worldwide basis. No less than 5 spacecrafts dedicated to the study of Comet Halley were launched during 1985, the rendezvous dates being all clustered around early March 1986, about one month after the comet's closest approach to the sun.

In the immediate run-up to these events Fred and I met to discuss what observations might be likely according to our present point of view. What predictions might we possibly make? Our deliberations led us to conclude that organic/biologic comets of the kind we envisage would have exceedingly black surfaces. This is due to the development of a highly porous crust of polymerised organic particles that can permit vigorous outgassing only when the crust comes to ruptured. We put all our arguments in the form of a preprint entitled "Some Predictions on the Nature of Comet Halley" dated 1 March 1986 (Cardiff Series, No 121) which came to be published much later in *Earth, Moon and Planets* (**36**, 289–293, 1986). This was only twelve days before the encounter, and our priority would have gone unrecorded had it not been for the fortunate circumstance that the *London Times* picked up on it and reported its contents (*The Times*, March 12, 1986).

On the night of March 13, 1986 we watched our television screens with nervous anticipation as Giotto's cameras began to approach within 500 km of the comet's nucleus. The fears that the spacecraft might be badly damaged and even destroyed by impacts with cometary dust were proved to be wrong, and the equipment functioned well throughout the encounter. The cameras were expecting to photograph a bright snowfield scene on the nucleus consistent with the then fashionable Whipple dirty snowball model of comets. In the event the television pictures transmitted world-wide on 13 March proved to be a disappointment. The cameras had their apertures shut down to a minimum and trained to find the brightest spot in the field. As a consequence, very little of any interest was immediately captured on camera. The much publicised Giotto images of the nucleus of Comet Halley were obtained only after a great deal of image processing. The stark conclusion to be drawn from the Giotto imaging was the revelation of a cometary nucleus that was amazingly black. It was described at the time as being "blacker than the blackest coal ... the lowest albedo of any surface in the solar system ...." Naturally we jumped for joy! As far as we were aware at the time we were the only scientists who made a prediction of this kind, a prediction that was a natural consequence of our organic/biologic model of comets. Fred and I regarded this development as yet another decisive triumph of our point of view. More triumphs were soon to follow.

A few days after the Giotto rendezvous, infrared observations of the comet were made by Dayal Wickramasinghe and David Allen using the 154 inch Anglo-Australian Telescope (*IUA Circular* No. 4205, 1986). On March 31, 1986 they discovered a strong emission from heated organic dust over the 2 to 4 μm waveband. As noted earlier basic structures of organic molecules involving CH linkages absorb and emit radiation over the 3.3–3.5 μm infrared waveband, and for any assembly of complex organic molecules such as in a bacterium, this absorption is broad and takes on a highly distinctive

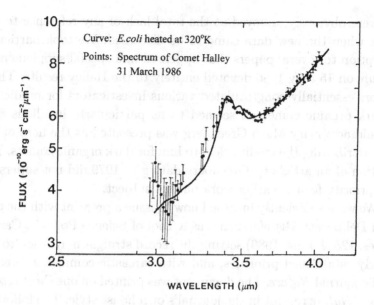

Fig. 15 Comparison of the spectrum of Comet Halley (points) obtained by Dayal Wickramasinghe and David Allen with a microbial model (curve).

profile. The Comet Halley observations by Dayal and David Allen were found to be identical to the expected behaviour of desiccated bacteria heated to 320 K (Fig. 15). Another victory for our model! Later analysis of data obtained from mass spectrometers aboard Giotto also showed a composition of the break-up fragments of dust as they struck the detector to be similar to bacterial degradation products.

The Halley observations, in our view, clearly disproved the fashionable Whipple's "dirty snowball" theory of comets. The theory dies hard, however, with variants of it still in vogue with the claim that Whipple was still mostly right, except that there was more dirt (organic dirt) than snow! It could not be denied that water existed in comets in the form of ice, but great quantities of organic particles indistinguishable from bacteria are embedded within the ice. This conclusion was unavoidable unless one chose to ignore the new facts. (D.T. Wickramasinghe, F. Hoyle, N.C. Wickramasinghe and S. Al-Mufti, *Astrophys. Space Sci.* **36**, 295–299, 1986). With such clear-cut documentation of our priorities for organic comet models

we were obviously annoyed at the total lack of any reference to our
work when the new data came to be discussed. We took particular
exception to several papers that appeared in a special supplement of
Nature on 15 May 1986 devoted entirely to the Halley results. These
papers essentially congratulated various investigators for predicting
a dark organic comet. It seemed to us particularly invidious that
our old adversary Mayo Greenberg was presented as the hero of the
day, attributing the credit solely to him for dark organic comets. The
citation of an article by Greenberg dating to 1979 did not supersede
our priority for the earlier work on this subject.

We were sufficiently incensed now to issue a preprint with the title
"On Deliberate Misreferencing as a Tool of Science Policy" (Cardiff
Series, 125, 1 June, 1986) setting the record straight in relation to our
clearly established priorities, and with sarcastic comments directed
at the journal *Nature*. The document was printed on one sheet resembling a *Nature* reprint in the journal's own house style. The following
quotes would suffice to convey the gist of our contention.

> *"Despite our admiration of Professor Greenberg's sibylline
> achievements, we think that most people require predictions to
> be announced publicly in advance of events. Otherwise there
> would be no point at all in making predictions. No ordinary
> bookmaker, for example, pays out money to would-be clients
> who seek to back a winner two months after a race has been
> run. In science, however, bookmakers gladly pay out money
> even years after a race has been run, especially it seems if the
> client happens to be Professor Greenberg.*
>
> *In the special supplement to the issue of Nature for 15 May
> there are two examples, both concerning the likely organic
> nature of the bulk of cometary dust. The communication
> "Composition of Comet Halley dust Particles from Giotto
> observations" (J. Kissel et al, p. 336) attributes the sugges-
> tion that cometary dust might be organic to an article by
> Greenberg dated 1979, while the communication "Composi-
> tion of Comet Halley from dust particles from Vega obser-
> vations" also references Greenberg, through an article dated
> 1982. Our own first paper on organic material appeared in*

*1975.... Our views by 1979 had indeed become so widely
known as to justify the word "notorious" ...*

*Misrepresentation would not be in question if our views
from 1975 onwards have been wholly wrong. It is just
because we have sometimes been right that misreferencing has
arisen....*

*Our opposition to the Darwinian theory lies, we believe, at
the root of the matter. At a microbial level, where only single base-pair changes of DNA are involved, the Darwinian
theory is correct. But at the macrolevel involving multiple
base-pair changes, the theory is wrong... Evolution that is
Darwinian proceeds mostly at the lowest level of all, the level
of varieties, just as it was conceived to do already in the
1850's. Beyond what was then apparent, the theory does not
do much. But even if all this was not clearly demonstrable
it would be usual in science, in physical science at any rate,
for us to be granted the right to our own opinion, instead of
being exposed to an orchestrated campaign of calumny and
misreferencing...."*

*There is no law that misreferencing shall not be used as
an instrument of science policy. Nor is there a law outside
the courts which even requires even an approximation to the
truth to be spoken or written... Yet it should be remembered
that cultures of high quality are more fragile than they appear
at first sight. While they were still operative there seemed no
limit to the productions of the Elizabethan dramatists, the
Florentine painters and the Viennese musicians. But all are
gone now and no amount of effort, desire or money will bring
them back. There is no more likely cause of a similar decline
and virtual disappearance of productive science it seems to us
than a calculated disrespect for the truth...."*

Such provocative words were not ignored by *Nature*. A response by
John Maddox appeared in the issue of *Nature* dated 19 June 1986
in the form of an article entitled "When reference means deference".
Referring to the direction that our work on grains had taken, he

opined thus:

> "*Many readers will know the eccentric direction this work has taken. Hoyle and Wickramasinghe have remarked that carbon is a common constituent of interstellar dust (as it is of the gas), that much of this carbon is in the form of sedimentary organic molecules and have then gone on, in a series of mostly samizdat publications, to argue that the prevalence of interstellar carbon only goes to support one of the other theories to which they are attached. They say that life as it is known on the surface of the Earth is genetically far too complicated to have arisen by means of Darwinian evolution from primordial chemicals, that at least the first steps in the evolution of life are likely to have taken place "out there" and that, as it happens, comets are likely to have been among the means by which the surface of the Earth is from time to time repopulated by external organisms ...*
>
> *Much of what they have written over the years deserved to be widely read, for example the argument that comets contain polyformaldehyde ... or the demonstration that comets Cernis and Bowell probably contain organic molecules rather than water ice. What these authors seem incapable of understanding is that their panspermian convictions sully even their sober contributions to the literature.*"

We later discovered, in going through old issues of *Nature* of the early years of the 20th century, that similar pronouncements about panspermian ideas were in vogue as early as 1890. The first tentative steps towards panspermia were taken by John Tyndall on January 21, 1890 at the Royal Institution in London. At a Friday evening discourse Tyndall demonstrated, by means of a simple optical experiment, the presence of dust in the atmosphere, and went on to make the bold assertion that such invisible dust could include a component that he called "vibriones" — germs that were carried in the air and caused epidemics of disease. *Nature* pounced on Tyndall with vengeance, saying that his imprudent speculation could open the floodgates to undesirable panspermian doctrines. A tirade of attack continued for weeks in the columns of *Nature*, but neither Tyndall, or his contemporary Lord Kelvin were deflected in their advocacy of panspermia. The events of 1986 showed us that nothing had changed in close to a century!

When Fred protested about Maddox's exposition, *Nature* invited us to write an article presenting our views. It was published with the title "The case for life as a cosmic phenomenon" (*Nature* **322**, 509–511, 1986).

Chapter 19

Alternative Cosmologies

I should mention that the academic year 1987/1988 turned out to be traumatic for reasons unconnected with our work. University College Cardiff was deemed to be in financial difficulty by the Government and Principal Bill Bevan was forced to resign. Furthermore, University College Cardiff and the neighbouring University of Wales Institute of Science and Technology was forced to form a merged University institution that came into legal existence in 1988 under the name University of Wales College of Cardiff, later to be called Cardiff University. In the process of merger individual departments in the two constituent colleges had also to merge. There were 4 mathematics departments at University College Cardiff and one at the University of Wales Institute of Science and Technology that were combined into a single School of Mathematics. The large contingent of astronomers in my old Department of Applied Mathematics and Astronomy were now "displaced persons" and were forced to join a Department of Physics, that later became the Department of Physics and Astronomy. I had a choice: I could have gone over to physics or stayed with mathematics. Fred advised me to stay in the School of Mathematics, which I did. His argument was that there would always be a call for mathematics, whereas if proposed new courses in astronomy somehow failed to catch on, my position in the new outfit may not be so secure.

So after 1988 my "group" within the School of Mathematics had dwindled to four: myself, Fred Hoyle, Max Wallis, Shriwan Al-Mufti and 2 research students. I also now ceased to be Head of Department, through a manoeuvre that was scarcely legal, with the result that I was given an unprecedented workload of teaching. Fred of course still continued to come to Cardiff from time to time. Needless to say I received much encouragement from his visits, particularly to feel that we were confronting the really important problems of the Universe, against the backdrop of which my own squabbles with the University could be viewed in proper perspective. Fred's visits to Cardiff in the late 1980's and early 1990's tended to be more relaxed than before. We felt now that most of the hard work had been done. A conceptual framework for a grand theory of cosmic life was fully in place, and its predictions were being borne out in observations from several disciplines. Interstellar dust and cometary dust were found to possess exactly the properties we had predicted, and the oldest life on the Earth was pushed back to a time when intense cometary bombardment was taking place. We had the strongest indication that comets seeded the planet with life some 4 billion years ago. The discoveries of microbial life enduring the most extreme conditions were suggesting an alien context for such properties, and opening possibilities of microbial habitats in a wide range of bodies in the solar system. A wrong theory does not come up repeatedly with such an indefatigable series of successes. Sooner or later a contradiction turns up and the theory has to be abandoned. This has not happened in our case. Why then is there such deep-rooted hostility to our ideas? Perhaps such ideas went against the grain of an essentially geocentric scientific culture?

We talked at length about the possible causes of the opposition we faced, often despairing for future generations to whom the search for objective truth would become an ever more distant goal. Whilst the real source of the antagonism remained a mystery, the lack of support from our home institution Cardiff was undoubtedly a contributory factor. Following the departure of Bevan and the dissolution of my department, it had become fashionable to dismiss our ideas as insane, and the message gradually filtered out to the outside world.

A major shift in attitude was apparent in other University institutions as well. A golden age when philosophers and academics held sway had given way to an age of hard headed accountants. Money was all that mattered. The search for truth was subjugated by an over-powering greed for accumulating research funds and political power. Nor were these unsavoury developments confined to the cloisters of academia. Indeed it appeared that Universities were merely respond-ing to major changes that were taking place in the world at large. Margaret Thatcher, whose government introduced the idea of a mar-ket economy for universities, was now in her third term of office as Prime Minister. People were reckoned to be less important than dubi-ously conceived objectives within institutions, including Universities. Intolerance of all sorts was on the rise. In 1989 Salman Rushdie pub-lished his novel *The Satanic Verses*, and the Iranian leader Ayatol-lah Khomeini ordered muslims to execute him. In Sri Lanka conflicts between Tamil separatists and the Government flared sporadically. In China racial attacks on black students early in 1989 were followed by the historic protest and massacre of dissidents in Tiananmen Square. Even in the UK the incidence of racial attacks were noticeably on the increase, with the Police and authorities often turning a blind eye. My wife and I continued to have our fair share of "go home Paki" comments in Cardiff, and similar written threats were delivered to me at the University by persons who could not be identified. I felt that I had suddenly become a victim of racial abuse and discrimination within my own University, and this was sometimes heartbreaking.

Despite all these problems Fred and I continued to pursue our own researches in whichever direction that new data directed. And new data did indeed come our way at a brisk pace. A discovery of a 3.28 μm emission feature in the diffuse radiation emitted by the Galaxy confirmed that aromatic molecules of some kind were exceed-ingly common on a galactic scale. We argued that the infrared emis-sions not just at 3.28 μm but over discrete set of wavelengths — 3.28, 6.2, 7.7, 8.6, 11.3 μm — must arise from the absorption of ultra-violet starlight by the same molecular system that degrades this energy into the infrared. We had shown much earlier that the 2175 Å extinction of starlight may be due to biological aromatic molecules, and it seemed natural then to connect the two phenomena. Thus we developed a unified theory of infrared emission and ultraviolet extinction by the same ensemble of aromatic molecules. A currently fashionable non-biological aromatic molecule was coronene ($C_{24}H_{12}$), and it was easy to demonstrate that this type of molecule was nowhere near as good as biological aromatics. (F. Hoyle and N.C. Wickramasinghe, *Astrophys. Space Sci.* **154**, 143–147, 1989; N.C. Wickramasinghe, F. Hoyle and T. Al-Jubory, *Astrophys. Space Sci.* **158**, 135–140, 1989.)

Returning to our earlier interest in exploring possibilities for the cosmic microwave background, we turned our attention next to iron whiskers. It is known that the element iron is produced in supernovae, and we could easily show that in an expanding supernova envelope containing newly synthesised nuclides ^{56}Ni and ^{56}Fe, iron particles would eventually condense from the gas when the temperature fell to 1000 K. Since laboratory studies showed that iron vapour has a tendency to condense into slender long whiskers, we argued now that a similar process would occur in supernovae. Iron whiskers of diameter $0.02\,\mu m$ and lengths of the order of 1 mm formed in this way are expelled into space. We computed the absorption properties of these whiskers using standard formulae and showed that they could have higher opacities at mm wavelengths than in the visual waveband. This is of course an excellent condition for modelling the cosmic microwave background — the light from distant galaxies will not be blocked, while microwave thermalisation can occur. We discovered also that iron whiskers will experience very high radiation pressure forces and could be expelled from galaxies and clusters of galaxies at high speed. (F. Hoyle and N.C. Wickramasinghe, *Astrophys. Space Sci.* **147**, 245–256, 1988). Extragalactic iron whiskers are just what we needed to rescue the beleaguered Steady State Cosmology.

Fred and I both attended the 22nd ESLAB Symposium that was held in the delightful Spanish town of Salamanca between 7 and 9 December 1988. We presented three joint papers here, two on our theory of interstellar grains and one on a model of the cosmic microwave background based on long iron whiskers. Apart from the presence of the predictably hostile Greenberg contingent, we felt there was a mellowing of attitude towards us, compared to our experiences both in Colombo and at the RAS a few years earlier. There was still a long way yet to conceding a biological grain model, but people were at least willing to listen to the arguments and to extract what they thought was its acceptable essence. Organic dust everywhere in the cosmos, including in the comets, had become the order of the day.

Our new microwave background theory might have provoked hostility, but it did not. Again peopled listened with tacit interest. In fact many of the participants used dictaphones to record our lectures. The paper describing the new model was published in the conference proceedings (F. Hoyle and N.C. Wickramasinghe, *ESA-SP*-290 489–495, 1989). It was fortunate that the new supernova SN187A came to be discussed at this meeting, and the first signs of dust condensation were becoming evident (W.P.S. Meikle, *ESA-SP*-290 329–337, 1989). Firm evidence in favour of iron whiskers around SN1987A was, however, to become available only much later (N.C. Wickramasinghe and A.N. Wickramasinghe, *Astrophys. Space Sci.* **200**, 145–150, 1993).

In the world outside, momentous changes were under way. Two such changes are worthy of note. The Berlin Wall that was erected in August 1961, when East Germany sealed off the border between East and West Berlin, was triumphantly and dramatically brought down in November 1989. The ugliest symbol of a divided Europe disappeared overnight and with it the Cold War had come to an end. The map of Europe had to be redrawn.

And in South Africa, even more spectacular events were taking place. For over half a century the white South African government, under the control of the Afrikaner National Party, had pursued a policy of apartheid, which involved a total racial segregation — interracial marriages and mixed-race sporting events being outlawed. In the 1970's the government had established tribal "homelands" in the poorest parts of the country. Blacks needed passes to work outside these "homelands". In most instances, only single persons or married men received passes so that when workers left the homelands, they had to leave their families behind. Racially segregated worker "townships" became established outside all the major cities, and the blacks lived under conditions of unspeakable squalor, police brutality and discrimination. The situation was economically so favourable for the White settlers that the prospect of change seemed only a distant dream in the mid 1980's. In 1989, under the Presidency of

F.W. de Klerk, changes came swiftly and unexpectedly. Although he knew he was going against the mainstream wishes of the South African whites, growing international pressure pushed de Klerk to work towards dismantling apartheid. And against all the odds he succeeded. In 1990, he released Nelson Mandela from prison, after over a quarter of century of incarceration, and started negotiating with him and the African National Congress (ANC) on the transfer of political power.

Quite a different kind of development that attracted our attention in the winter of 1989 was an outbreak of influenza throughout the UK. It was described as the worst influenza epidemic for 12 years, and hospitals were inundated with cases of complications from the flu. Fred and I decide to undertake our second Hoyle–Wickramasinghe Influenza Survey by sending out questionnaires to all independent schools in the UK, and by visiting particular schools in our immediate neighbourhood. Priya and I went in person to discover what happened in a place not far from Cardiff. We discovered that among the earliest to succumb in the new outbreak were the inhabitants of the sleepy little village of Gowerton, near Swansea. Attendance registers at the local school showed a sudden rise of absences on 27 November, the same day when a local publican, his wife and their child all came down with flu. Furthermore it was documented that the entire village became shrouded in a persistent low-lying mist starting about two days before the outbreak.

Although the influenza type known as H3N2 was the dominant subtype found to be involved in the epidemic, a whole host of other airborne viruses appears to have been in circulation at the same time. The incidence patterns of these viruses were consistent with an atmospheric fall out model, and inconsistent with direct person-to-person spread. All the data we collected appeared to corroborate our findings from the 1977/78 pandemic — even the epidemic patterns at Eton College. At this stage we were surprised to receive an invitation from the Royal Society of Medicine to write a review paper of our findings about influenza. Our article appeared as: "Influenza — evidence against contagion: discussion paper" (F. Hoyle and N.C. Wickramasinghe, *J. Roy. Soc. Med.* **83**, 258, 1990).

Our work in areas relating to cosmic life continued for the next half decade at least. Apart from presentations at international conferences and numerous public lectures that we gave, we had also begun writing a technical monograph, *The Theory of Cosmic Grains* (Kluwer Academic Publishers, 1991). We also started to think more explicitly about the relationship of life to cosmology. The Universe had to be of such a kind that the superastronomically improbable origin of life occurred. The use of the existence of life as an indicator of a hitherto unknown property of the Universe is known nowadays as the Weak Anthropic Principle. This principle was pioneered by Fred in the 1950's by examining a condition that is needed to produce adequate amounts of the elements carbon and oxygen in stars. I have already mentioned that to achieve this Fred deduced that the nucleus of ^{12}C must possess an excited state close to 7.65 MeV above ground level. No such state was known to exist at the time when this deduction was made, but a state with exactly the predicted energy was discovered shortly afterwards.

In order to overcome the superastronomical information hurdle of life (to which I have referred earlier), Fred began to turn again to Steady-State Cosmology. In our preprint dated March 1991 "The Universe and Life: Deductions from the Weak Anthropic Principle" which was published later in *Astrophysics and Space Science* (F. Hoyle and N.C. Wickramasinghe, *Astrophys. Space Sci.* **265**, 89–102, 1999), we argued that access to superastronomical masses of carbonaceous matter is to be needed to get life started in the Universe. This leads to a sharp distinction between Big Bang and Steady-State cosmologies. In the standard Big Bang cosmology the total mass of carbonaceous matter available for life is a mere 10^{40} g, and the time available is limited to 15–18 billion years. In a steady-state or quasi-steady-state universe, the situation is dramatically different. In a quasi-steady-state model of the universe discussed by Fred in collaboration with Geoff Burbidge and Jayant Narlikar (F. Hoyle, G. Burbidge and J.V. Narlikar, *A different approach to cosmology*, Cambridge University Press, 2000), the universe expands exponentially on a timescale of 1000 billion years, while undergoing oscillations on a timescale of 50 billion years which continue for an

eternity. New matter is created at the beginning of each oscillation
when the density of the universe is highest. It is this matter that is
processed in stars during each cycle and turned into carbonaceous
material that is available for biological processing. If one starts with
a biological message in some lifeless part of such a universe at a particular place at a particular time, and if the message can be copied
and distributed, then in the Quasi-Steady-State Cosmology we estimated that after a hundred billion Earth-ages, the message would
have spread through $10^{90,000,000}$ grammes of carbonaceous material.
This means that there is now a superastronomical mass of carbonaceous matter within which the superastronomically improbable event
of the origin of life could occur. Thus life may arise in a non-living universe in about a hundred billion Earth-ages, provided the Universe is
in a steady-state or a quasi-steady-state. All this suggests that Fred's
thoughts were never far divorced from cosmology, even when we were
pondering the seemingly different problem of the origin of life.

In the summer of 1989 Jayant Narlikar contacted me to ask if
I might be able to host a small cosmology workshop in which Fred
could discuss the current state of cosmology with his closest collaborators. We first considered Gregynog Hall in mid-Wales as a possible
venue for the meeting but decided against it because of the logistics
of transporting participants to and from airports. Between 25 and
29 September 1989, Fred Hoyle, Geoffrey Burbidge, Harlton Arp,
Jayant Narlikar and I met at a small residential conference centre
— Dyffryn Gardens — just outside Cardiff city. We discussed the
many lines of evidence that all appeared to go against the standard
Big Bang model of the universe.

By now the evidence accumulated by Harlton Arp and Geoff Burbidge concerning QSO's with large redshifts being physically linked
to galaxies of low redshift seemed to put paid to the assumption that
high redshifts imply cosmological distances. The insecure nature of
many of the other assumptions in the conventional Big Bang theory were also discussed and led this group to propose an alternative
cosmology that was more consistent with the facts. My own contribution was mostly directed towards interpreting the cosmic microwave
background. This has been the strongest piece of evidence that had

been cited in support of a hot Big Bang universe for over two decades. At the time of the Cardiff meeting new measurements made by the COBE satellite were pointing to a microwave background that had a black body spectrum to a very high degree of approximation, and an isotropy on angular scales down to a few arc minutes. All this was being further adduced as support for the Big Bang cosmological model. Fred was quick to point out that the new COBE observations posed a problem for the Big Bang models. An early Black Body spectrum must surely come to be distorted by subsequent events — the condensation of galaxies and clusters of galaxies in the primordial universe must surely leave a mark on the isotropy of the background.

Our alternative point of view was that the cosmic microwave background was the final end product of the thermalisation of the energy produced by the conversion of hydrogen to helium in stars. The thermalisation is a multi-stage process in our view: starlight energy is first converted into the infrared by the normal dust in galaxies, then the infrared in turn is degraded into microwaves by absorption and re-emission by millimetre long iron whiskers.

Our deliberations at the meeting were written up in the form of an article and sent to John Maddox, Editor of *Nature*. It is still a little bewildering how Maddox was persuaded to publish such a devastating attack on conventional cosmology. The paper appeared with the title "The Extragalactic Universe: an alternative view" in the issue of *Nature* of 30 August 1990 (H.C. Arp., G. Burbidge, F. Hoyle and N.C. Wickramasinghe, *Nature* **346**, 807–812, 1990). Referring to the implied fallacy of the conventional explanation of the cosmic microwave background we wrote thus:

> "*The commonsense inference from the Planckian nature of the spectrum of the microwave background and from the smoothness of the background is that, so far as microwaves are concerned, we are living in a fog and that fog is relatively local. A man who falls asleep on the top of a mountain and who wakes in a fog does not think he is looking at the origin of the Universe. He thinks he is in a fog.*"

Our alternative view of the cosmic microwave background could not have been put more succinctly.

Chapter 20

The Last Decade

We did not stall in any of our various projects during the decade 1990–2000, nor did we feel we had come to the end of the road. Barbara's health was giving cause for concern throughout this period so Fred was finding it increasingly difficult to spend time away from home. Home for the Hoyles was now in Bournemouth, where they had moved from the challenging climes of the Lake District, partly owing to Barbara's health. This meant we saw less of him in Cardiff although our interaction and collaborations continued, by telephone and fax.

Fred's book, *Ice*, published in 1981 (F. Hoyle: *Ice*, Hutchinson & Co. Lond, 1981), had introduced the question of ice ages being mediated by cometary dusting. To examine this question more thoroughly we needed an accurate determination of the behaviour of ice particles in the stratosphere. The requirement now was to calculate the fraction of incident sunlight that was absorbed and scattered in directions that did not reach the Earth. This part of the incident sunlight is therefore lost to the process of heating the Earth. I performed the relevant mathematical analysis for the geometry of the Earth using standard theories of light scattering by spherical grains, and sent this to Fred in the autumn of 1990. Fred checked every step of my calculation and our joint paper "Back-scattering of sunlight by ice grains in the mesosphere" was published in *Earth, Moon and Planets* (F. Hoyle and N.C. Wickramasinghe, *Earth, Moon and Planets* **52**, 161–170, 1991), and paved the way to other related projects. Several shorter contributions from us followed, all of which stressed the sensitivity of the Earth's climate to stratospheric particle loading, particles that were derived from either terrestrial or extraterrestrial sources (e.g., Hoyle and Wickramasinghe, *Nature* **350**, 467, 1991).

Throughout much of the decade we continued to pursue various consequences of stratospheric dusting. In 1996, we collaborated with Bill Napier and Victor Clube to explore the effects of the break up of a giant comet, leading to fragments in Earth-crossing orbits, and to recurrent episodes of impacts and cometary dusting on the Earth. We argued for a connection between such cometary events, periodic glaciations as well as episodes of mass-extinctions in the geological record. We also suggested that the entire history of human civilization, over the past 10,000 years, after the end of the last ice age, bears witness to a record of repeated episodes of assaults from the skies (S.V.M. Clube, F. Hoyle, W.M. Napier and N.C. Wickramasinghe, *Astrophys. Space Sci.* **245**, 43–60, 1996). In a later paper, Fred and I also worked out a more specific analysis of the possible connection between cometary events and ice ages (F. Hoyle and N.C. Wickramasinghe, *Astrophys. Space Sci.* **275**, 367–376, 2001).

The modelling of ice ages took us in a somewhat esoteric direction to examine more carefully an idea that Fred had touched upon in a preprint some years earlier. The connection might sound bizarre, but our studies of ice ages led us to the origin of race prejudice among humans. This issue had come recently to the fore in view of a much publicised report of a British government inquiry into the gratuitous murder of Stephen Lawrence (a black youngster) in South East London, in April 1993. The London Metropolitan Police had refused to prosecute the white youths who murdered Lawrence, and the family launched a private prosecution that led to a Government inquiry under the chairmanship of Sir William Macpherson. The Macpherson report, published in March 1999, found the Metropolitan Police to be "institutionally racist", a phenomenon that was said to exist also in other institutions.

How, one might ask, could such a strong emotional response to skin colour prevail at a time in our civilization when we pride ourselves as being "enlightened"? When Fred and I first discussed this, we soon agreed that there must be a powerful biological imperative for racism to persist, not just in Britain, but throughout much of the modern world. We published our speculations on this subject in 1999 in the *Journal of Scientific Exploration* (**13**, 681–684, 1999). The theory is very simple, if somewhat Lamarckian in its implications.

It is an undeniable fact that human evolution over the past 2 million years has led to the emergence of two broadly distinct groups of people with regard to skin colour, one fair the other dark. The lighter skinned group now occupies countries in northern latitudes that were on the borders of glaciers during ice ages, and the darker skinned group mainly inhabit temperate and equatorial regions. The difference in skin colour between these groups hinges on a variable efficiency to produce the pigment melanin. Melanin is produced in a special group of cells known as the melanocytes that are located at the base of the skin, and although the density of such cells remains more or less invariable, the efficiency of melanin expression is highly variable. Many genes appear to be involved in melanin production, and the overall situation for melanin expression (that is to say for being black) is strongly dominant.

The present-day situation for maintaining selective pressures for melanin expression rests on a razor's edge between two competing effects. On the one hand melanin as a pigment protects the base of the skin from being damaged by ultraviolet radiation from the Sun which in extreme instances leads to carcinomas of the skin. On the other hand, an adequate penetration through the skin of ultraviolet radiation with wavelength shortward of 3130 Å is needed for the production of Vitamin D. Whilst the latter requirement is not too relevant in the present day with high levels of nutrition generally, it would have been a strong selective factor for survival in harsher prehistoric times. With lower levels of dietary acquisition of Vitamin D, the lack of an adequate absorption of sunlight would lead to the crippling disease of rickets. In this disease severe bone deformities result from the lack of Vitamin D, a substance that plays a crucial role in the absorption of calcium from food. The correct level of pigment expression depends on the available ultraviolet light at any given location at a given time. The balance is between rickets causing skeletal deformities with a consequent lower fecundity, and excessive sunburn radiation leading to skin cancers with attendant high levels of mortality.

The higher incidence of skin cancers in lighter skinned Caucasian migrants in the tropics is well attested. Likewise a high incidence of rickets has been recorded in black and Asian populations living in northern latitude countries before the large-scale introduction of vitamin supplements into staple foods. At the beginning of the 20th century, rickets was reported to affect 90% of black infants in New York. Even as recently as the 1970's high rates of incidence of rickets have been recorded in children of Asian immigrants living in Britain, as for instance in a Glasgow based study. The intensity of sunlight available to Asian immigrants in countries like the UK is obviously mismatched to their expressed pigment level, but routine food fortifications and vitamin pills now generally compensate for this deficiency. Needless to say such dietary supplements were unavailable in prehistoric times.

Throughout the Pleistocene epoch when human evolution occurred, the Earth was locked in an ice age that lasted for nearly 2.5 million years. There were warmer interglacial remissions interspersed throughout this time, each one lasting for about 10,000 years, and the total duration of all such warm periods making up just 10 per cent of the entire Pleistocene era. The Earth emerged from the last ice age approximately 11,000 years ago.

During ice ages the average temperature of the Earth's surface was about 10 degrees Celsius colder than today, and ice sheets were about three times as extensive as they are now. White skinned Nordic tribes living close to the edge of rugged windswept ice sheets under grey skies would have been eking out a precarious existence, grabbing whatever food could be gathered and utilising every photon of ultraviolet from the sun in order to stay alive and free of rickets. For them survival was crucially contingent upon having their genes for melanin suppressed. For people living in the tropics, however, the drier ice-age conditions with less cloud cover than at present would have made for a remorseless flood of ultraviolet radiation to fall on their skins. Survival for them was contingent on the fullest expression of their melanin genes, being as black they possibly could be.

In the course of random migrations from the south, the white populations living at the edge of ice sheets would have been at risk through matings with people possessing darker skins. Black–white matings would have tended to produce offspring with darker skins and thus more prone to rickets. Fewer of these malformed children would reach reproductive age, so black–white matings posed a real extinction threat to the white races. Under such circumstances the emergence of mating prohibitions and colour prejudice would be a natural outcome. The prejudice would become deeply ingrained in social traditions, language, mythology and religion. Thus the depiction of Satan as a black figure cannot be regarded as accidental, nor can the association of evil generally with blackness. The strong emotions manifest in modern racism could be understood, but not forgiven, in these terms.

Returning from this extended diversion back to our mainstream activities, we examined more carefully the theory behind our earlier computations of cross sections for iron whiskers. It would be recalled that these calculations were crucial for non-cosmological explanations of the cosmic microwave background. One deficiency of our earlier work that began to worry us was an assumption we made relating to the electrical conductivity of iron at low temperatures and low frequencies. We had taken a frequency-independent value of conductivity of 10^{18} s^{-1} somewhat arbitrarily. Was there a better way to deal with this matter? In attempting to answer this question we came upon a way of using a well-attested theory of metals known as the Drude theory to work out the low temperature dielectric function of iron as a function of frequency. The only input that was now required was the DC conductivity of iron which of course was well known. Our calculations which broke new ground were published in 1994 (Wickramasinghe and Hoyle, *Astrophys. Space Sci.* **213**, 143–154, 1994) and the formulae set out here were used in all our subsequent calculations involving iron whiskers.

The dust in interstellar space continued to turn up many surprises even after more than three decades. Keeping abreast with the latest developments in this field, we noticed a remarkable observation that had gone almost unnoticed. Dust in a wide range of astronomical situations — reflection nebulae, planetary nebulae, HII region, high latitude galactic cirrus and in the extended halo of the external galaxy M82 all showed a broad emission feature over the waveband 6000–7000 Å. Astronomers were trying to identify this feature as being an effect of the so-called PAH molecules in interstellar space. But the fits for inorganic PAH's that had been considered in relation to this data remained poor. We discovered that a much closer fit could be obtained if one considered a fluorescence phenomenon that is well-known for many biological pigments. A galaxy like M82 appeared to be glowing in a pigment similar to that present in glowworms! (F. Hoyle and N.C. Wickramasinghe, *Astrophys. Space Sci.* **235**, 343–347, 1996).

The giant Comet Hale-Bopp made its début in the Spring of 1997. It was certainly the brightest comet that had been seen for some time. The comet was estimated to have a nucleus of some 40 km in diameter and an orbital period, as it came in from the outskirts of the solar system, of about 4200 years. It brightened steeply as it came into perihelion on April 1 with a tail extending over 10–30 degrees, and remained a spectacular object for most of March and April 1997. Many types of molecules including organic molecules were discovered spectroscopically in the coma of Comet Hale-Bopp, and the infrared spectrum over the wavelength range 2.5–45 μm range was measured by the European Space Agency's Infrared Space Observatory (ISO) satellite when the comet was at a distance of 2.9 Astronomical Units from the sun. This spectrum is displayed by the jagged curve of Fig. 16. The dashed curve shows a spectrum we calculated for a model involving approximately 90% by mass of a bio-culture including diatoms, and 10% in the form of pure olivine dust. Although an inorganic olivine component is needed to fit the positions of the peaks near 10 μm, this alone is hopeless for explaining the full range of the data. A dominant contribution from an organic (biologic) component is required, and any larger contribution from olivine than about 10 percent is inconsistent with the data. We published our results first in the new internet journal *Natural Science*

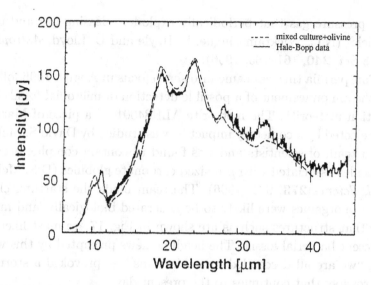

Fig. 16   Fit of the spectrum of Comet Hale-Bopp to a mixture of microorganisms and crystalline olivine.

of May 1997, and later in *Astrophysics and Space Science* (**268**, 379–383, 1999).

A further indication of cometary biology was the discovery that comets show considerable activity when they are at relatively great distances from the Sun, far beyond the orbit of Jupiter. This was observed for comet Halley after its 1986 perihelion when it had retreated to a distance of some 6–10 Astronomical Units from the Sun. Similar sporadic outbursts of activity had been known to be occurring in the case of comet Schwassmann–Wachman I, which has a period of about 15 years and an orbit that lies outside the boundaries of Jupiter and Saturn. Now in the case of the new comet Hale-Bopp it was reported that before it came to perihelion (in August, September and October of 1995), it was producing extensive dust and carbon monoxide halos. We analysed this data and concluded that a strong case for biological outgassing could be made. Inorganic comets would not be expected to explode in the manner that Comet Hale-Bopp had done in the cold depths of space. But biological activity simmering beneath a frozen crust and intermittently provoked by meteorite impacts could lead to the build-up pockets of high pressure gas that periodically explode, releasing gas and dust particles (N.C. Wickramasinghe, F. Hoyle and D. Lloyd, *Astrophys. Space Sci.* **240**, 161–165, 1996).

Panspermia theories came into sharp focus in August 1996 following the announcement of a possible detection of microbial fossils in a Martian meteorite. The meteorite ALH84001 — a piece of Martian rock ejected by a cometary impact — was studied by David S. McKay and a team of scientists and was found to contain complex organic molecules associated with μm-sized carbonate globules (D.S. McKay *et al, Science* **273**, 924, 1996). The team made the startling claim that the organics were likely to be generated biologically, and moreover that structures such as are shown in Fig. 17 are most likely to represent bacterial fossils. The headline news prompted by this work that "we are all descended from Martians" — provoked a storm of controversy that continues to the present day.

Although that claim itself has since been challenged, the impact of the initial announcement has not diminished in the intervening years. Astrobiology has suddenly emerged as a new scientific discipline and

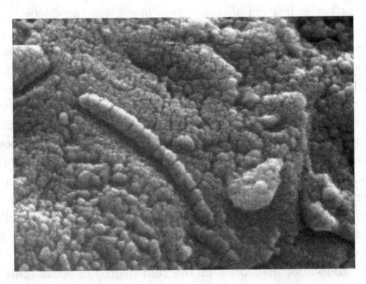

Fig. 17   Putative microbial fossils in the Martian meteorite ALH84001 (Courtesy NASA).

several international organisations including NASA have expressed their commitment to research in this general area. Connecting an impending paradigm shift in a roundabout way to Mars was a decision that was politically astute. For the concept of life on Mars has been filtering slowly into the public's consciousness since at least 1898 when H.G. Well's novel, *The War of the Worlds* was first published and introduced the frightening fiction that Martians were threatening to invade the Earth.

The Mars meteorite ALH84001 has shown beyond any doubt that complex organic structures, and by inference even microbial cells, could be transferred in a viable form from one planetary body to another. Planetary panspermia or transpermia as this concept has recently come to be known, is not by any means a new theory. It was discussed by Lord Kelvin well over a century ago. In his presidential address to the 1881 meeting of the British Association, Kelvin drew the following remarkable picture:

> "*When two great masses come into collision in space, it is certain that a large part of each is melted, but it seems also quite certain that in many cases a large quantity of debris must be shot forth in all directions, much of which may have experienced no greater violence than individual pieces of rock experience in a landslip or in blasting by gunpowder. Should the time when this earth comes into collision with another body, comparable in dimensions to itself, be when it is still clothed as at present with vegetation, many great and small fragments carrying seeds of living plants and animals would undoubtedly be scattered through space. Hence, and because we all confidently believe that there are at present, and have been from time immemorial, many worlds of life besides our own, we must regard it as probable in the highest degree that there are countless seed-bearing meteoric stones moving about through space. If at the present instant no life existed upon the earth, one such stone falling upon it might, by what we blindly call natural causes, lead to its becoming covered with vegetation.*"

Thus the ideas that have recently come to the forefront of scientific discussion were in circulation over 123 years ago. Such interplanetary transfers of life as described by Kelvin are possible of course, but in our view they represent a relatively unimportant route for exchange of life on a cosmic scale. Moreover they fail to address the all-important question of how life began in the solar system in the first place. According to the ideas we have discussed earlier, a far better option is to have Mars, Earth and every other habitable planetary body infected with the same cometary source of life — a source of life that is derived from an even bigger system. As we have already noted, comets impact all planetary bodies, and so cometary panspermia must surely remain the principal route for the transference of cosmic life.

It has been widely claimed that interstellar panspermia is unlikely because of the hazards of ultraviolet light and ionising radiation that has to be faced by iterant bacteria (C. Mileikowsky, *et al, Icarus* **145**, 391, 2000). A typical transit time between amplification sites in the galaxy could take a few million years, and the fraction of survivors needed for our theory to be valid is less than 10^{-22}. Firstly, we have argued that ultraviolet radiation is very easily protected against: only a thin layer of carbonaceous coating around a bacterium provides almost complete shielding. Ionising radiation could pose a bigger threat, but survivors are still inevitable, at least in the cases where entire comets could be transported from one planetary system harbouring life to another nascent system. Even individual bacteria or clumps of bacteria could withstand the doses of ionising radiation received during transit. Experiments suggesting otherwise are based on large fluxes of ionising radiation delivered in seconds or minutes, whereas in the interstellar medium a trickle of such radiation is incident over millions of years. The two situations could be dramatically different and not directly analogous as is normally assumed. It is well known that the oxidising effects of free radicals, particularly the hydroxyl radical, cause over 90% of DNA damage. So reducing the water content (from which hydroxyl is derived) can drastically diminish the lethal effects of ionising radiation. Moreover, irradiation in an inert atmosphere or vacuum such as exist in interstellar space, would also reduce potential damage. Low temperatures also go in the same direction by immobilising and preventing the diffusion of free radicals. For all these reasons it is fair to surmise that there is a considerable uncertainty as to the effects of cosmic radiation on interplanetary or interstellar bacteria. A low flux of ionising radiation delivered over astronomical timescales to dormant freeze-dried bacteria (in the absence of H_2O and air) would perhaps bear no comparison with equivalent doses on vegetative cultures in the laboratory (N.C. Wickramasinghe and J.T. Wickramasinghe, *Astrophys. Space Sci.* **286**, 453, 2003).

Direct proof of the survival of bacteria exposed to radiation environments in the near Earth environment has also been demonstrated using NASA's Long Exposure facility (G. Hornek, *et al, Adv. Space Res.* **14**, 41, 1994). Viable cultures of bacteria have been recovered from ice drills going back 500,000 years, from isolates in amber over 25–40 million years (R.J. Cano and M. Borucki, *Science* **268**, 1060, 1995) and from 120 million year old material (C.L. Greenblatt *et al, Microbial Ecology* **38**, 58, 1999). Similarly viable bacteria were recovered in salt crystals from a New Mexico salt mine dated at 250 million years (R.H. Vreeland, W.D. Rosenzweig and D. Powers, *Nature* **407**, 897–900, 2000). The present day dose rate of ionising radiation on the Earth arising from natural radioactivity is in the range 0.1–1 rad per year. Well-attested recoveries of dormant bacteria/spores after 100 million years imply tolerance to ionising radiation with total doses in the range ~10–100 million rads. All the indications are that a large enough fraction survives to ensure the operation of panspermia, even for "naked" bacteria or bacterial clumps.

The theory that we developed throughout our journey requires life to have been introduced to Earth for the first time by comets some 4 billion years ago. But that process could not have stopped at a distant time in the past. Comets are still with us, and the Earth is entwined in the debris shed by comets. We know that at the present time some 100 tonnes of cometary material reaches our planet on a daily basis. One might then ask: What evidence is there of living particles, microbes, coming in with this influx of debris? Much of the infalling cometary debris would of course be in the form of millimetre-sized or larger particles that burn up as meteors on entry. But a significant fraction of infalling cometary material will be of sizes that will enable them to travel safely through the atmosphere, and this, according to our ideas, must include clumps of bacteria, including nanobacteria and viruses, freshly released from cometary surfaces.

I have previously discussed our theory of the 1980s that bombardments from space could lead to pathogenic interactions with higher life forms. And our interest in this process never ceased. In December 2000, during the epidemic of BSE that was raging through farms in the UK, we wrote the following letter to the *Independent*:

"THE CAUSE OF BSE
SIR — Diseases of plants and animals have a long history of mysterious appearances, like mysterious characters that appear inexplicably on stage in a play, without any satisfactory explanation being offered as to where they have come from. An example some years ago was the lethal respiratory disease that suddenly hit the grey seals in the remote Siberian Lake Baikal.

As the remarkable complexity of genetic systems comes increasingly to light it should be obvious that life on the Earth is far too intricate to have evolved here in isolation from the rest of the Universe. It is because life here is a part of a far vaster system that it is so complex.

The connection comes in our view from material of cometary origin being incident on the Earth in considerable amount. Recent studies have shown that much of the material of escaping from comets is in the form of organic particles that cannot be distinguished from biomaterial. The mass input to the Earth is estimated to be several tens of tonnes of cometary stuff per day, sufficient if it was all in the form of bacteria to give a daily incidence of several hundred thousand bacteria per square metre of area. For the most part the material proves to be harmless. It simply washes away. But in rare cases a connection may occur and if a connection escalates, mostly due to fortuitous circumstances, a new disease is born."

Small particles of bacterial and viral sizes descend through the Earth's stratosphere mostly during the winter months, and in our opinion it was the nearly unique English practise of out-wintering cattle that explains why BSE hit English farms more severely than elsewhere. English farmers move cattle frequently from field to field, maximising their chance of picking up any pathogen that may fall from the air onto the grass.

Once a causative agent (genetic fragment or piece of infective protein) got into a few cattle man took a hand, by grinding up infected animals and including them in feed for more cattle. In retrospect this may look a foolish thing to have done, but without knowing what was going on it is roughly comprehensible on economic grounds.

We live nowadays in a blame culture, egged on relentlessly by television. Somebody, we are constantly being told, has to be held responsible for BSE, when according to our point of view there was no culprit, not unless blame be equated with ignorance. Indeed the political authorities, by banning the inclusion of infected portions of cattle in cattle feed, may be said to have acted both quickly and responsibly.

Whether they should also have banned any use of cattle products in medical vaccines remains another question with disturbing possibilities.

Prof. Sir Fred Hoyle
Prof. Chandra Wickramasinghe"

In February 1999 the Stardust Mission to Comet Wild 2 was launched with the aim of conducting *in situ* experiments as well as collecting samples of cometary dust. The rendezvous and collection took place in January 2004. The collection was executed using an aerogel block to gently break the speed of falling cometary particles, but even so it was not expected that any microbes could survive the impact on the aerogel. When the material was finally returned to Earth in 2006, fragments of organic structures bearing tell-tale signs of life were all that we were able to find.

The most promising method of detecting incoming cometary microorganisms is the use of sterile collection systems sent on balloons into the high stratosphere. Above the tropopause, which is 18 km in the tropics and 10 km in temperature latitudes, aerosols of 1–10 μm in size, including bacterial clumps, could not stay for more than a very short timescale, weeks or less. They would quickly fall under gravity. If small amounts of bacteria from the Earth's surface get lofted on rare occasions to great heights, for example after a volcanic eruption, they would quickly fall. Above 40 km you would not expect to find any terrestrial bacteria at all in normal times, so if significant quantities of stratospheric bacteria are discovered, this would provide *prima facie* evidence of panspermia.

For many years Fred and I tried to convince organisations that had the capacity to carry out such an experiment that this was a project of potential value. The responses we received were consistently discouraging except in one instance. Fred had visited Jayant Narlikar at the Tata Institute on several occasions during the 1980's and had suggested that they attempted to collect cometary dust in the stratosphere using sterile equipment carried on balloons. The Tata Institute Balloon launching facility had a distinguished track record since the 1950's when they were engaged in pioneering work on the detection of cosmic rays. Evidently, Fred was given a polite hearing by the Indian scientists but the general impression at the time was that the experiment was not feasible because collection procedures that were sufficiently aseptic were not available. Methods of detecting small amounts of DNA, using amplification techniques for example, were also not developed at the time. By the late 1990's the situation had significantly changed. Under the leadership of Jayant Narlikar a team of physicists and biologists proposed an experiment of the kind suggested by Fred Hoyle and myself to the Indian Space Research Organisation (ISRO). Funding was finally approved in the year 2000.

The object of the experiment was to collect stratospheric air aseptically, and to examine it in the laboratory for signs of life. The collection part of the project was as follows. A number of specially manufactured sterilized stainless steel cylinders were evacuated to almost zero pressures and fitted with valves that could be open and shut on ground telecommand. An assembly of such cylinders was suspended in a chamber of liquid neon to keep them at cryogenic temperatures, and the entire payload was launched from the TATA Institute Balloon launching facility in Hyderabad, India on 21 January, 2001. As the valves of the cylinders were opened at predetermined heights, ambient air rushed in to fill the vacuum, building up very high pressures within the cylinders. The valves were shut after a prescribed length of time, the cylinders hermetically sealed and parachuted down to the ground.

A set of cylinders was transported to Cardiff in February 2001, but due to the bureaucratic difficulties I had been experiencing with the authorities at Cardiff University the samples were not extracted from them until April 2001. Fred was kept fully appraised of developments of the balloon experiment and he gave us valuable advice at every stage. Priya and I saw Fred for the last time on 21 February 2001 in Bournemouth. The first question he asked me was "What have you found in the balloon samples?" My answer was that we were still waiting to appoint an assistant to do the work — a situation he found difficult to comprehend.

Eventually I managed to secure a minimal extent of support from Cardiff University and the short-lived services of a research assistant. The cylinders were opened and the collected stratospheric air made to flow through sterile membrane filters in a contaminant free environment. Any bacteria or clumps of bacteria present in the stratospheric air sample would then be collected on these filters. The analysis was first conducted for us by the microbiology department in Cardiff, and later investigations at Sheffield University were led by Milton Wainwright.

The first phase of this investigation was completed in July 2001 and we had unambiguous evidence for the presence of clumps of living cells in air samples from as high as 41 kilometres, well above the local tropopause (16 km), above which no micron sized aerosols from lower down would normally be expected to transported. The detection was made using electron microscope images (see Fig. 18), and by using

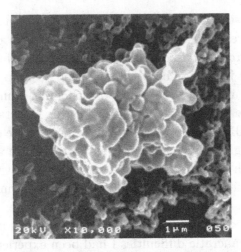

Fig. 18 A cluster of putative microorganisms collected from the stratosphere imaged using a scanning electron microscope.

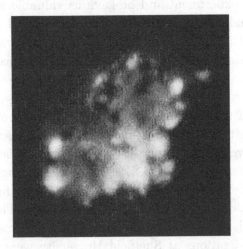

Fig. 19 A cluster of viable microorganisms from 41 kilometres, identified using a fluorescent dye taken up only by live cells.

a fluorescent dye known as Cyanine that is only taken up by the membranes of living cells. When the isolate treated with the dye was examined under a special kind of microscope the picture on the Fig. 19 was obtained. DNA was also detected in these clumps of cells using yet another fluorescence technique.

The variation with height of the density of such cells indicated strongly that the clumps of bacterial cells are falling from space. The input of such biological material was provisionally estimated to be between 1/3 tonne to 1 tonne per day over the entire planet. If this amount of organic material was in the form of bacteria, the annual transfer of bacteria is 10^{21}. These results were presented by me on behalf of our team at the Instruments, Methods, and Missions for Astrobiology IV session of the SPIE Meeting in San Diego at the end of July 2001. The paper was entitled: "The detection of living cells in stratospheric samples" by Melanie J. Harris, N.C. Wickramasinghe, David Lloyd, J.V. Narlikar, P. Rajaratnam, M.P. Turner, S. Al-Mufti, M.K. Wallis, S. Ramadurai and F. Hoyle, *Proc SPIE* **4495**, 192, 2002. This is sadly the last paper that Fred co-authored. The presentation made international headlines. Doubts about contamination were naturally raised by sceptics with a geocentric worldview, but our initial results have since received extensive confirmation in the work carried out by Milton Wainwright in Sheffield (M. Wainwright, N.C. Wickramasinghe, J.V. Narlikar and P. Rajaratnam, *FEMS Microbiology Letters*, **218**, 161, 2003; M. Wainwright, N.C. Wickramasinghe, J.V. Narlikar, P. Rajaratnam and J. Perkins, *Int. J. Astrobiol*, **3**, 13, 2004).

Science is certainly progressing towards vindicating the point of view that Fred Hoyle and I had developed over several decades. I have recently explored an alternative route to panspermia which compliments our previous work. (W.M. Napier, *Mon. not. R. Astron. Soc.* **348**, 46, 2004; M.K. Wallis and N.C. Wickramasinghe, *Mon. not. R. Astron. Soc.* **348**, 52, 2004.) Just as comets and asteroids impacting the Earth, after it was populated with life, could lead to extinctions of species, they could also splash material laden with life back into space. A fraction of this life-bearing material survives the ejection process and can actually escape from the solar system. The solar system (including the Earth) revolves around the centre of the Galaxy once in 240 million years, and the life-bearing material ejected from Earth would have periodic access at close quarters to hundreds of millions of nascent cometary and planetary systems. Such newly forming planetary systems would then be recipients of Earth life in the form

of viable microorganisms. And since the Earth cannot be the sole centre of life, this same dissemination process would happen for every
other life-bearing planetary system.

Space exploration of comets could be said to have barely got
under way in the year 2004. From the time of the Giotto probe of
Halley's comet in 1986, standard dogmas about comets continued
to be revised. And the trend has been unerringly in the direction
from inorganic comets to organic and life-bearing comets, just as
we had first proposed in 1979. The most recent Rosetta mission to
comet 67P/Churyumov-Gerasimenko (of which the author is a science team member) was launched in February 2004 with a landing
on the comet's surface scheduled for 2014. The latest data, however,
came from the Stardust Spacecraft's encounter with Comet Wild 2
which took place in January 2004 at a distance from the comet of
just 236 kilometres (147 miles). The Principal Investigator of the
Stardust Mission, Donald Brownlee is reported as saying:

> *"We thought Comet Wild 2 would be like a dirty, black, fluffy
> snowball. Instead it was mind-boggling to see the diverse land-
> scape in the first pictures — including spires, pits and craters,
> which must be supported by a cohesive surface . . . "*

Stardust images of Comet Wild 2 showed pinnacles towering to
heights of 100 metres and craters plunging to depths of more than 150
metres. The entire comet is only about five kilometres across, yet one
of the largest craters is itself a kilometre across. Comet Wild 2 was
imaged at close quarters (Fig. 20) and showed features that could
not be further removed from the old "dirty snowball model", one
that had acquired a hallowed status in cometary theories of the 20th
century. Most puzzling of all are the dozens of extended jets — particularly jets that emanated from the unlit dark side of the comet
that faced away from the sun. Comet Wild 2 has provided startling
evidence of a seething hot cauldron of organic material bubbling
beneath a frozen crust. Weak spots on the crust become ruptured
from time to time, venting the high pressure fluid beneath thus giving
rise to jets. The new data is strikingly consistent with our biological

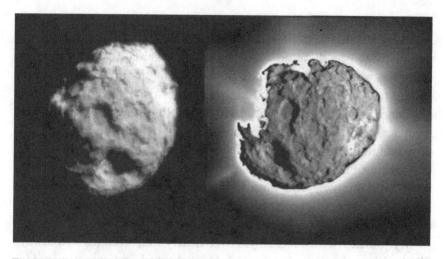

Fig. 20   Images of Comet Wild 2 taken in January 2004 with cameras aboard STARDUST (Courtesy, NASA, JPL).

model of a comet. Evidence supporting cosmic life seems now to be inescapable.

Over half a millennium ago, the Polish astronomer Nicolaus Copernicus (1473–1543) dethroned the Earth from its privileged position at the centre of the cosmos. The Earth, however, continued, well into the 20th century, to occupy an exalted status as the centre of life. Life was regarded as the result of terrestrial evolutionary processes occurring independently of the external Universe. Throughout our journey, Fred and I have sought to challenge this position using evidence derived from many different fields of science. Life on Earth cannot be regarded as being isolated from the rest of the galaxy. We are part of a truly enormous cosmic gene pool. Nothing of great innovative significance in biology ever happened on the Earth. The Earth was simply a receiving station, a building site for the incomparably magnificent edifice of cosmic life. When prejudice against the concept of cosmic life has run its course, our ideas will come to be regarded as self-evident truths. The cosmic quality of life will seem as obvious to future generations as the Sun being at the centre of our solar system is obvious to us today.

The sad news of Fred Hoyle's death at the age of 86 on 21st August 2001 marked the end of an era of 20th century science, a pursuit that was dictated solely by intellectual curiosity to discover what the world was really like. Sociological constraints were subjugated and the man-made boundaries between the various disciplines of science were broken. As Fred Hoyle would often say, the Universe itself has no respect whatsoever for such artificial boundaries. His own monumental work on nucleogenesis in the 1940's and 1950's, introducing nuclear physics to astronomy, epitomised this principle. So also did our own joint work that I have described in this book that lifted the barriers between biology and astronomy and gave birth to the burgeoning new science of astrobiology.

My journey with Fred Hoyle over nearly four decades was always filled with action — adventures into uncharted and sometimes dangerous terrain, and the excitement of new discoveries. If I were given another chance, I would gladly follow the same path again.

Where our journey ends
Another must begin
In some distant corner of the Universe
Following inexorably
The selfsame path.

Reflections in 2012

"Except for a very few scientists, everybody overlooked a crucial step in the analogy between commercial and natural selection. Commercial selection works only because at the back of it there are human intellects constantly striving to improve the range and quality of their products. Commercial selection is therefore very far from the purposeless affair natural selection is taken to be in biology.

In reality, natural selection acts like a sieve. It can distinguish between species presented to it, but it cannot decide what species shall be sieved in the first place. The control over what is presented to the sieve has to enter terrestrial biology from outside itself — not just from outside the living world, but from outside the confines of our planet.... There is nowadays a mountain of evidence for this view.

Fred Hoyle, The Intelligent Universe, 1983
(Michael Joseph, London)

Search for the Origin of Life

Since the first edition of this book was published there have been many scientific developments that could have tested our thesis that life is a cosmic phenomenon. The fact that every new discovery turned out to support rather than contradict the theory gives added confidence to the belief that we are on the right track towards unravelling our cosmic ancestry.

Following the successes of Harold Urey and Stanley Miller in the 1950's research into the origins of life continues stridently in many laboratories in the hope that a breakthrough will ultimately be achieved in understanding how non-living organic matter turns into life. Spectacular successes in genetic engineering and gene manipulation have been achieved, including the wholesale reconstruction of a genome inside a DNA-free bacterial cell (Gibson *et al.*, 2010). But all this is still a far cry from understanding the origin of life itself. What we may conclude from our current knowledge of biochemistry and microbiology in 2012 is not significantly different from what we had in the early 1980's.

There is no evidence whatsoever that requires life to have started *de novo* on the Earth. Indeed all the evidence from geology and astronomy now points inexorably to an origin of life that lies well outside our planet, exactly as Fred Hoyle and I had first proposed in 1979–1981. The picture emerging is of an initial injection of a viable cellular life form, that takes root on our planet and thereafter begins to evolve, being genetically augmented from time to time by genes of external origin that enable the development of new kingdoms of life, phyla, genus and species. What seemed to be an outrageous heresy 30 years ago is at last coming to be accepted, albeit grudgingly.

Cost of Heterodoxy

The history of science has many examples of innovators whose ideas were so far ahead of their time that they failed to gain acceptance. Such individuals often suffered cruel penalties. Anaxoragas (500–428BC) famously argued that the Sun was a red hot stone, and the Moon was made of earth, and for this impiety he was banished from Athens. Giordano Bruno (1548–1600), who maintained that the universe was full of inhabited planets, was condemned for heresy and burnt to death. In our more civilised world we at least pay lip service to the libertarian principles of toleration of opinion and attempt to encourage a diversity of views. So any ostracism or obstacles Fred and I faced in our own journey must be reckoned to be mild compared to what had gone before! As I pointed out in earlier chapters our work was often seen to be destitute of reverence for authority, and for this reason shunned. However, to have one's work ignored is infinitely better than suffering persecution, including the burning of one's books that had happened in past ages.

Our experience over several decades brought to light a dangerous modern trend in the sociology of 20th and 21st century science. There is a tendency nowadays for authorities wielding power to withhold support or recognition of work that does not conform with orthodox opinion. I have indicated how this had happened at various points in our story.

The idea that the organic building blocks of life were brought to Earth by comets rather than being synthesised in the atmosphere, was first developed by us in our book *Lifecloud* in 1976. Although this was vigorously resisted at the outset, the idea came to be slowly accepted by the scientific community because it was thought to fall within the general paradigm of the "primordial soup on Earth", even though the soup itself had to be imported from space.

Throughout the 20th century and in the first decade of the present century the vastness of the Universe of galaxies, stars and planets has been reaffirmed by every major astronomical breakthrough. And with developments in biochemistry and microbiology the bewildering complexity of molecular arrangements in even the simplest living cell have all pointed to a cosmic rather than a terrestrial origin of life.

It was known for a long time that certain types of microbes possessed properties that were not obviously related to the "average" conditions that prevail on Earth. Research into the properties and distribution of "extremophiles" is now being claimed as evidence that life can indeed survive in the harshest of extraterrestrial environments, and that the transfer of life from one galactic location to another was entirely feasible. In 2010 it was discovered that cyanobacteria placed on the outside of the International Space Station survived alternations of freezing, heating and exposure to harsh ionising radiation for a full 18 months. Scarcely a decade ago the very existence of such microorganisms might have been thought impossible.

Astronomical Predictions: Comets and Meteorites

Our theory of the cosmic origins of life made a prediction that the infrared and ultraviolet spectral properties of interstellar dust would match biological material, which it convincingly did when the first such spectra were obtained. We further predicted that the dust from comets (hitherto thought to be made of inorganic ices) must in fact be organic, and this prediction too was verified by observations of comet Halley and other comets after 1986.

I referred in Chapter 20 to air samples that may contain comet dust being collected aseptically by the Indian Space Research Organisation (ISRO) from a height of 41 km in the stratosphere and showing evidence of microorganisms of a presumed cometary origin falling over the whole Earth at an average rate of 0.1 tonnes per day (Wainwright *et al.*, 2003). A few years later, in a second stratospheric sampling, three new bacterial species with exceptional ultraviolet resistance properties were isolated, and one of these was named in honour of Fred Hoyle — *Janibacter hoylei* (Shivaji *et al.*, 2009). The decision to name a new bacterium after Hoyle would appear to have been made in recognition of his pioneering contributions to astrobiology, in the same way that naming a main-belt asteroid 8077 Hoyle (1986AW2) was in recognition of his longstanding contributions to astronomy.

The importance of having similar experiments conducted by other independent Space Agencies cannot be overemphasised, particularly in view of the enormous importance of firmly establishing the cometary origin of these organisms. The cost effectiveness of such a project is beyond dispute, and the apparent reluctance to conduct these experiments is, in my view, connected with the fear of decisively overturning the long-held paradigm of Earth-centred life. Significantly no life detection experiment has been included in any of the recent space missions, again reflecting a state of mind of a scientific orthodoxy unwilling to countenance the non-terrestrial origin of life.

In 2006 samples of dust from comet Wild 2, secured in NASA's Stardust Mission, were returned safely to Earth. As expected, the high velocities of impact onto the collecting aerogel blocks left little evidence of any original organic grains or putative cells — only trails of molecular debris. Whilst no living cells were recovered, complex organic molecules were found in abundance in the debris trails, including an amino acid; and all this was consistent with the break-up of biological material. The biological explanation for the genesis of this material is by far more plausible than the claim that the organics may represent products of radiation processing of simpler molecules. In addition to organics, the collected material contained mineral particles, including the mineral known as cubite. Since cubite can only be formed in the presence of liquid water this discovery provides dramatic confirmation of our prediction from the 1970's that liquid water existed in primordial comets and played a crucial role in the replication of cometary bacteria.

Direct evidence of water jets from a comet was discovered in 2011 in a photograph of comet Tempel 1 taken by cameras onboard the same spacecraft that conducted a rendezvous with comet Wild 2 in 2004. These water jets could result from fissures in the comet's crust due to the build-up of gas pressure from bacterial metabolic activity,

Fig. 21   Braided water jets from comet Tempel-1 photographed in Feb 2011 by *Stardust* spacecraft.

thus leading to the release also of gas and dust (Wickramasinghe, Hoyle and Lloyd, 1996).

NASA's Deep Impact Mission to Comet Tempel 1 on July 4th 2005 involved a high-speed impact of a probe at 37,000 km/hr onto the comet's surface, rupturing its crust and releasing and exposing its content. The presence of water and complex organics that could be connected life and mineral dust was revealed, although the unambiguous discovery of living cells was precluded by the nature of the experiments (Wickramasinghe, Wickramasinghe, Napier, 2010).

The European Space Agency's ROSETTA mission to comet 67P/ Churyumov-Gerasimenko (on which I am one of a very large team of investigators) was launched in March 2004. It is due to attempt a landing on the comet's surface in November 2014 and carry out a wide range of experiments that could give us a deeper insight into the structure, the chemistry and physics of comets.

It was pointed out in Chapter 13 that a class of meteorite known as carbonaceous chondrites represent relics of comets that had once contained microbial life, and thus we might expect to find fossilized microorganisms in such meteorites (Figure 13). I have already referred to the pioneering work of Hans Pflug in discovering such microfossils in the 1980's using the best equipment available at the time and with the most stringent precautions to avoid contamination. Although Pflug's work did not receive the attention it deserved, being apparently so far ahead of his time, the same programme of work continued at NASA's Marshall Space Flight Centre under the direction of Richard Hoover. Hoover has now published a vast catalogue of fossilized microbial structures which he identifies with various types of microorganisms (Hoover, 2005, 2011). An example of what he finds in a freshly cleaved surface of the Murchison meteorite is given in the SEM (scanning electron microscope) image shown in Fig. 22. The comparison (left frame) is with a modern specimen of living cyanobacteria. Many arguments to support the claim that such organic structures in the meteorite are indigenous to the meteorite and not contaminants have been given. The meteorite structures show chemical signatures (e.g. low nitrogen content) to indicate that they are indeed fossilized structures, and therefore cannot be modern contaminants.

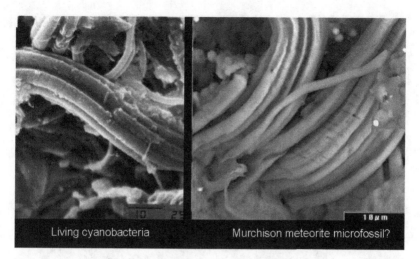

Living cyanobacteria | Murchison meteorite microfossil?

Fig. 22   A structure in the Murchison meteorite (Hoover, 2005) compared with living cyanobacteria (Hoover, 2005, 2011).

Studies of the Mars meteorite ALH84001 that were discussed in Chapter 21 have continued to yield results that are consistent with the original claim of detecting microbial fossils. In July 2011 another meteorite from Mars fell over the desserts of Morocco and was recovered soon afterwards in October near the village of Tissint. This so-called *Tissint* meteorite was blasted off the surface of Mars by a comet or asteroid impact several million years ago. A piece of this meteorite was recently examined by my PhD student Jamie Wallis and other collaborators and we reported the discovery of "signs of extinct life" in this meteorite as well (Wallis *et al.*, 2012).

Spherical globules rich in carbon and oxygen were discovered in the interior of the meteorite embedded in its rocky matrix. Figure 23 shows one such apparently hollow structure that cracked like an egg when subjected to a high energy electron beam.

Astronomical Spectroscopy

It is exactly 50 years since we published our paper on carbon grains in the Universe (Hoyle and Wickramasinghe, 1962). It is this original work that led to the development of the organic theory of interstellar

Fig. 23   A carbon-oxygen rich particle in the interior of the Tissint meteorite: An ovoid shaped carbon-oxygen rich globule cracking under gold coating in a scanning electron microscope examination (Wallis *et al.*, 2012).

dust, and eventually the theory of cosmic life. The correspondences of the predictions of our model with astronomical observations, discussed earlier in this book, continue to be verified as the quality of data has improved with the use of new telescopes such as the Spitzer Space Telescope. The high resolution infrared spectra of galactic and extragalactic sources show spectral features of dust ("unidentified infrared bands", UIB's, principally at 3.3, 6.2, 7.7, 8.2 and 11.3 micrometres) that can only reasonably be interpreted as biologically generated heteroaromatic molecules (Wickramasinghe, 2010; Rauf and Wickramasinghe, 2011). Non-biological explanations that are on offer are contrived and, moreover are inconsistent with *all* the findings.

Amongst the most distant galaxies displaying aromatic/biomolecular infrared signatures is a high red-shift infrared luminous galaxy at redshift $z = 2.69$, the spectrum of which is shown in Fig. 24 (Teplitz *et al.*, 2007). This galaxy emitted its light when the Universe was at the tender age of 2 billion years according to standard Big Bang

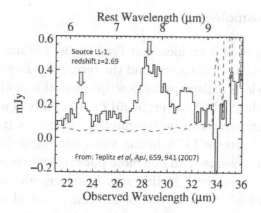

Fig. 24 Redshifted 6.2, 8.7, 11.3 micron bands in the source (Teplitz *et al.*, 2007).

cosmology. Another spectral signature in the ultraviolet (absorption at 2175 Å) pointing to biochemicals has also been found in galaxies at similar great distances (Elíasdóttir *et al.*, 2009; Motta *et al.*, 2002; Noterdaeme *et al.*, 2009).

The idea that the material causing such emissions represents the degradation products of biology, as we had originally suggested, and continue to stress, can be challenged only on the grounds that extraterrestrial life is an "extraordinary hypothesis" requiring "extraordinary evidence" to support it. The confinement of life to the Earth is indeed the more extraordinary hypothesis, and far less justifiable since mechanisms of viable transfer of microbes across the galaxy have been clearly identified (Wickramasinghe *et al.*, 2010).

Biological molecules may well have existed even much earlier in the history of the universe, but the discovery of such molecules is still to come. Evidence of dust extinction has indeed been found in very distant galaxies, and a high concentration of the life-element carbon has recently been detected in the most distant radio galaxy at a redshift of $z = 5.19$ (Matsuoka *et al.*, 2011). If the existence of life is judged by carbon abundance, then we can infer that the first signs of life appear within a billion years after the Big Bang — ready for cometary panspermia thereafter.

Big Bang Cosmology

The Big Bang itself is an idea that Fred Hoyle continued to question throughout much of his career, and the term *"Big Bang"* was in fact coined by Hoyle as a disparaging description of this model of the Universe that he disliked! However, in 2012 a vast body of modern observations in astronomy can be interpreted to imply a Big Bang-style origin of the Universe 13.75 billion years ago. Most (if not all) the matter we can observe with our most powerful telescopes originated in a gigantic "explosion" in this way. However, whether this was a unique "creation" event or one of an infinite set of similar events (multiverse, oscillating universe or Quasi Steady State Cosmology) still remains open to question.

The 2011 Nobel Prize for Physics was awarded to Saul Perlmutter, Brian Schmidt and Adam Reiss for their work on measuring the redshifts (distances) of the faintest Type 1a supernovae in the Universe. This led to the conclusion that the universe was accelerating or speeding up in the rate of its expansion. If the universe is dominated by normal matter this will not be possible — gravity must inevitably slow down, not speed up, the expansion rate. To resolve the dilemma a repulsive source of "dark energy" was invoked to overcome gravity, and this led the so-called "standard concordance model" which is comprised of 23% dark energy, 73% dark matter and 4% normal matter. As I pointed out in Chapter 19 an origin of life will be difficult to envisage in any such Big Bang cosmology that contains only 10^{40} grammes of carbonaceous material — that too existing in a highly dilute and dispersed state.

A variant of the standard Big Bang cosmology that has recently interested me is the so-called Hydro Gravitational Dynamics (HGD) cosmology of Carl Gibson and Rudy Schild (Gibson, 1996; Gibson and Schild, 2010). In the HGD cosmology models, the ionised gas of the early universe becomes unstable at the epoch when recombination from ionised to neutral gas occurred some 0.3 million years after the Big Bang. The whole universe was essentially transformed at this time into a "sea" of Earth-mass planetary objects resembling giant comets — 10^{80} in all. A fraction of these planetary bodies collide and coalesce into rapidly evolving massive stars, in which the chemical elements of life are synthesised by nuclear reactions on a short timescale. Explosions of supernovae then disperse these chemical elements into the great mass of primordial planets, and it is within the warm watery interiors of these objects — forming a sort of cosmological primordial soup — that an origin of life would have the best chance to happen (Gibson *et al.*, 2010).

According to this type of HGD cosmology the dark matter in the Universe is made up of life-bearing primordial planets. In our galaxy such planets are located in the a gigantic halo that continually feeds the spiral arms with material out of which stars and planetary systems form. Microbial life that originated within the first million years of the history of the Universe remains deep frozen and dormant until it is distributed within comets of newly forming planetary systems, and thence transferred to habitable planets like the Earth.

Planets

The cosmic theory of life discussed in earlier chapters requires microbial life to colonise every habitable niche within our own solar system. I have already discussed evidence of microbes in comets, in comet dust that enters the stratosphere, and in the residues of comets in the form of carbonaceous meteorites. Moon rocks brought back to Earth many years ago show no signs of life because particles of life-bearing comet dust impact the moon's airless surface at speeds of over 10 kilometres per second, with no atmosphere to break the speed and ensure a soft landing. Such high impact speeds are enough to rupture the bonds of organic compounds, leading only to a mysterious excess of carbon in moon rock that has often been claimed.

Mars is most likely to be the first external planet where the detection of life will be confirmed. In view of recent discoveries of microbial life in the harshest environments on the Earth, the possibility of microbial life on Mars must be close to certain. A major goal of the space probes Viking 1 and Viking 2 that landed on Mars on 20 July and 3 September 1976 was in fact to search for microbial life. Indeed the discovery of life on Mars may have already been made in 1976 by

Gil Levin, a Principal Investigator on this mission, but regrettably went unnoticed. A recent thorough re-examination and reappraisal of all the 1976 Viking data has confirmed that the only viable explanation of the results obtained from the Viking probes is on the basis of extant life on Mars.

It is a sad commentary that in the many robotic missions that have since landed on Mars not a single life-detection experiment was included. There is a touch of irony in that in future sample-return missions to Mars, in which rock samples will be brought back to Earth, elaborate "planetary protection" measures are to be adopted to take care of the contingency that microorganisms might be brought back from Mars — even, perhaps, microbes that may be pathogenic to humans!

Since 1976 there been many space missions to Mars. They have found evidence of subsurface water, dried-up river beds as well as methane in the upper atmosphere, all of which suggest that microorganisms could still live in specialised niches close to the surface. And in the distant past, when rivers flowed on Mars, much more abundant life was possible.

NASA's Curiosity Rover equipped with the most sophisticated mobile laboratory landed on the Gale crater of Mars on 6 August 2012 and is billed to spend several years probing for signs of past and even present life — albeit indirectly. If extant Martian life is found it can only be regarded as a long overdue confirmation of a discovery already made by Levin in 1976.

Another project of considerable interest is the proposed *in situ* exploration of three of Jupiter's icy moons, Callisto, Europa and Ganymede, the funding for which was approved by ESA (the European Space Agency) in 2012. All three of these Jovian moons have tidally-heated subsurface liquid oceans, and so are likely homes for life. Although this mission will not be launched for another decade and would reach Jupiter only in 2030, the prospect of a major astrobiology breakthrough at this time remains a strong possibility.

The search for habitable planets outside our solar system has gathered momentum following the launch of NASA's Kepler Mission in 2009. This mission deploys a 0.95 metre orbiting telescope to detect planets using "the method of transits", a procedure that looks for periodic dips in a star's brightness as an orbiting planet comes in front to partially block out the light from the star. Over 1000 definite detections of extra-solar system planets (or exoplanets) have been confirmed up to August 2012. Due to an observationally determined selection bias, most of the planets so found happen to be gas giants (Jupiter-like or Neptune-like) in close orbits around their parent stars. However, a few planets have been detected that are similar to giant Earths orbiting their parent stars at a distance that would permit liquid water to exist at the surface, and therefore capable of supporting life.

Perhaps the most interesting recent exoplanet discovery is Kepler 22b, a planet orbiting a G-type (Sun-like) star (Kepler 22) located in the constellation of Cygnus about 600 light years away. This planet has a radius roughly 2.4 times that of the Earth and lies within the habitable zone of a Sun-like star (G-dwarf). Further searches are expected to show up many similar exoplanets in the near future, and current estimates of the total number of such alien Earths in the galaxy run into tens of thousands, within a radius of 1000 light years from the sun. With the deployment of larger space telescopes for exoplanet studies as are currently being planned, spectra of planets like Kepler 22b could be obtained in the near future. We may hope to find chemical fingerprints of gases like water vapour, oxygen, ozone, carbon dioxide and methane that could reveal signs of life.

Besides planets that are in orbit around parent stars there is also a great deal of evidence of free-floating interstellar planets, and their total number could exceed the number of stars by factors of thousands (Wickramasinghe *et al.*, 2012). Such interstellar planets would not have their surfaces heated by the radiation of parent stars, but they could have interior domains of water kept liquid and warm by radioactive heat sources.

Evolutionary Predictions

The predictions of the cosmic-life theory included those related specifically to biological evolution (Hoyle and Wickramasinghe, 1979, 1981). We argued that if comets brought the first life to Earth 4 billion years ago, the process of microbial additions from comets must have continued throughout geological time, and consequently played a role in evolution. Such considerations were later extended to a model where genetic products of local evolution on a planet like the Earth were distributed and mixed on a galactic scale. We argued that comet impacts, such as happened at the K/T boundary 65 million years ago leading to the extinction of the dinosaurs, causes the inevitable splash back into space of DNA fragments that carry the products of local evolution (Wallis and Wickramasinghe, 2004; Napier, 2004; Wickramasinghe *et al.*, 2010). Even partially destroyed DNA strands belonging to life-forms that evolved locally could carry the information of life far and wide (Wesson, 2011). In this model similar impact episodes and gene distribution events would happen recurrently whenever the cloud of comets surrounding our planetary system is disturbed by the gravitational effect of a passing interstellar cloud. We estimate the *average* time interval between successive impact episodes to be about 40 million years, so that from the time when life first appeared on Earth some one hundred such gene distribution events would have taken place (Wickramasinghe *et al.*, 2010). We estimate that genes from Earth would thus have infected millions of nascent planetary systems throughout the Milky Way.

Since we cannot consider the Earth or our own solar system to be unique in this regard, it has to be assumed that similar gene dissemination processes operate for every life-bearing planet in the galaxy. As a consequence, the biosphere in which Darwinian evolution occurs must extend beyond our solar system to encompass a large fraction of the volume of the Milky Way. The stochastic nature of gene acquisition events resulting from encounters with molecular clouds leads naturally to a stochastic component of biological evolution — e.g. sudden jumps, as is apparently observed in the Earth's record of life.

Explicit Predictions from 1982

In *Proofs that Life is Cosmic*, (Hoyle and Wickramasinghe, 1982) pp. 73, 74 we wrote as follows:

"If we had knowledge that evolution was an entirely terrestrial affair then of course it would be hard to see how viruses from outside the Earth could interact in an intimate way with terrestrially-evolved cells, but we have no such knowledge, and in the absence of knowledge all one can say is that viruses and evolution must go together. If viruses are incident from space then evolution must also be driven from space. How can this happen? Viruses do not always attack the cells they enter. Instead of taking over the genetic apparatus of the cell in order to replicate themselves, a viral particle may add itself placidly to one or other of the chromosomes. If this should happen for the sex cells of a species, mating between similarly infected individuals leads to a new genotype in their offspring, since the genes derived from the virus are copied together with the other genes whenever there is cell division during the growth of the offspring....

A gene that happens to be useful to the adaptation of one life-form may be useless to another. Incidence from space knows nothing of such a difference, however, the gene being as likely to be added to the one form as the other. So genes that become functional in some species may exist only as nonsense genes in other species. This again is true. Genes that are useful to some species are found as redundant genes in other species. Suppose a new gene or genes to become added to the genotype (genome) of a number of members of some species. Suppose also that one or more of the genes thus added could yield a protein or proteins that would be helpful to the adaptation of the species. The cells of those members of the species possessing the favourable new genes operate, however, in accordance with the previously existing genes, and thus a problem arises as to how the new genes are to be switched into operation so as to become helpful to the species.... As potentially favourable genes pile up more and more, a species acquires a growing potential for large advantageous change, it acquires the potential for a major evolutionary leap, thereby punctuating its otherwise continuing state of little change — its 'equilibrium'...."

The process of horizontal gene transfer (HGT), which is now amply documented (Keeling and Palmer, 2008; Boto, 2010), provided the commentary to this quotation. The cosmic theory of life *requires* that genes which are the products of evolution in some distant cosmic location (comets or planets) can, on occasion, be transferred to evolving lifeforms on the Earth (Hoyle and Wickramasinghe, 1979, 1982).

In this way evolutionary advantage or novelty could be acquired by terrestrial organisms on a stochastic basis, whenever alien genetic material carrying new information is introduced to the Earth and becomes accessible to terrestrial biology. We thus inadvertently proposed an astronomical process of horizontal gene transfers — transfer of genetic information across normal mating barriers on a cosmological scale — before it was firmly demonstrated to operate as a process within terrestrial biology.

There is now compelling evidence to support a once contentious view that HGT provides an important source of new genes and functions to recipient organisms and also a driving force for evolution. It has also been recognized that the operation of horizontal gene transfer has foiled attempts to reconstruct ancient phylogenetic relationships in the search for a Last Universal Common Ancestor (LUCA) in the tree of life (Jain *et al.*, 2003). It is becoming clear that there was probably no such entity localized on the Earth but rather a cosmic ensemble of genes that has an antiquity comparable perhaps with the age of the Universe itself (Joseph and Wickramasinghe, 2011; Gibson *et al.*, 2011).

From all the available data we can infer that sudden shifts in evolution, the emergence of new traits and even the arrival of new species occurs through horizontal gene transfers rather than by the slow neo-Darwinian process of mutations and natural selection (Keeling and Palmer, 2008). Although the occurrence of neo-Darwinian evolution is not denied, it would probably be dwarfed by horizontal gene transfers in the long term. The phenomenon described by biologists as "punctuated equilibrium", where long periods of evolutionary stagnation are punctuated by sharp episodes of innovation and progress, is consistent with cosmically mediated gene transfers. The long periods of slow evolution are due to Earth-bound neo-Darwinian processes where no external gene inputs occurred (Wickramasinghe, 2012).

Viral Sequences in Genomes

Sequencing the human genome has been one of the most remarkable scientific developments of the new millennium. It has led to a wide range of discoveries that are transforming our ideas about viruses, disease and evolution (Venter *et al.*, 2001). One surprise was that the number of genes in human DNA (sequences coding for proteins) was as small as 20,000–25,000 rather than over 100,000 as had hitherto been suspected. Another surprise was that 50% of our DNA consists of sequences ultimately attributable to viruses. The best documented sequences correspond to so-called endogenous retroviruses — RNA viruses that reverse transcribe their RNA into DNA — which make up 8% of our DNA. Their significance in causing disease as well as contributing to evolution is only just coming to be understood, and many astounding correspondences with our 1979–81 statements quoted earlier cannot be overlooked (Wickramasinghe, 2012).

The new evidence from genome sequence studies points to frequent episodes of retroviral infections (of which HIV is an example) not only in humans, but in almost all mammalian species. De Groot *et al.* (2002) have identified an entire repertoire of genes known as (MHC class 1 genes) in chimpanzees that confer immunity against chimpanzee-derived simian immune deficiency virus (similar to human HIV). The inference is that modern chimp populations represent descendents from the survivors of a HIV-like pandemic that very nearly culled the entire ancestral chimp line in the distant past. The Hoyle–Wickramasinghe contention that HIV was an invader from space was much ridiculed when we first suggested it, but recent developments would appear to restore it at least to the realm of reasonable hypothesis.

The process by which viruses are "endogenised" and included in host genomes is not confined to retroviruses. A non-retrovial RNA transcript appears to have been incorporated in the germ line of several mammalian species, including rodents around 40 million years ago (Horie *et al.*, 2010). Bacterial infection can also leave an imprint on genes. A recent paper has shown that two immunomodulatory genes (known as SIGLEC), related to bacterial infection, are inactive in humans, but not in related primates (Wang *et al.*, 2012). The conjecture is that these genes when they were fully active could have been targets for a lethal bacterial infection that nearly culled the human population in the past, perhaps 100,000 years ago.

In our writings in the 1980's and 1990's we suggested that it would be prudent to maintain a microbiological surveillance of the stratosphere in a search for incoming potential pathogens so that vaccines may be developed, if necessary, to avert a future pandemic (Hoyle, Wickramasinghe and Watkins, 1986; Hoyle and Wickramasinghe, 1990). We predicted that, in general, weeks to months would elapse between the introduction of viral particles at the top of the stratosphere and their descent to ground level. This would give enough time for action. The time may well be ripe to consider instituting planetary protection protocols for such a contingency, before a devastating pandemic provides macarbe proof the theory of cometary panspermia.

My journey that began in the serene tranquillity of the English Lake District in autumn of 1961 has, after more than half a century, led to where I now stand. The rustic charm of the Old Dungeon Ghyll Hotel and my fireside discussions with Fred on cosmic dust are now distant memories, and so also is my first experience of wearing mountaineering boots and clambering up the craggy peaks of Langdale. The journey described in this book in search of our cosmic origins led far beyond Langdale pikes, through wild, unchartered and treacherous country. We faced danger at every turn, and conflict with adversaries. But true to the spirit of the indomitable mountaineer we plodded on, resolutely overcoming obstacles, our sights set on a distant Utopia where the true nature of our cosmic origins will be revealed for all to see. Such a goal at last appears within reach, and our ideas that were once considered heretical are moving imperceptibly into the realms of orthodox science. Our cosmic ancestry can no longer be denied. Terrestrial life originated and evolved from cosmic bacteria, augmented by genes that were also of cosmic origin. But the precise manner and process by which non-living matter in Universe turned into life in the first instance may remain a problem that eludes us for generations to come.

Bibliography to First Edition

Technical Papers

1. "A note on the origin of the Sun's polar field", F. Hoyle and N.C. Wickramasinghe, *Mon. not. R. Astron. Soc.*, **123**, 51, 1962.
2. "On graphite particles as interstellar grains", F. Hoyle and N.C. Wickramasinghe, *Mon. not. R. Astron. Soc.*, **124**, 417, 1962.
3. "On the deficiency in the ultraviolet fluxes from early type stars", F. Hoyle and N.C. Wickramasinghe, *Mon. not. R. Astron. Soc.*, **126**, 401, 1963.
4. "Impurities in interstellar grains", F. Hoyle and N.C. Wickramasinghe, *Nature*, **214**, 969, 1967.
5. "Condensation of the planets", F. Hoyle and N.C. Wickramasinghe, *Nature*, **217**, 415, 1968.
6. "Solid hydrogen and the microwave background", F. Hoyle, N.C. Wickramasinghe and V.C. Reddish, *Nature*, **218**, 1124, 1968.
7. "Condensation of dust in galactic explosions", F. Hoyle and N.C. Wickramasinghe, *Nature*, **218**, 1127, 1968.
8. "Interstellar grains", F. Hoyle and N.C. Wickramasinghe, *Nature*, **223**, 459, 1969.
9. "Dust in supernova explosions", F. Hoyle and N.C. Wickramasinghe, *Nature*, **226**, 62, 1970.
10. "Radio waves from grains in HII regions", F. Hoyle and N.C. Wickramasinghe, *Nature*, **227**, 473, 1970.
11. "Primitive grain clumps and organic compounds in carbonaceous chondrites", F. Hoyle and N.C. Wickramasinghe, *Nature*, **264**, 45, 1976.
12. "Organic molecules in interstellar dust: a possible spectral signature at 2200 Å?", N.C. Wickramasinghe, F. Hoyle and K. Nandy, *Astrophys. Space Sci.*, **47**, L1, 1977.

13. "Polysaccharides and the infrared spectrum of OH26.5 + 0.6", F. Hoyle and N.C. Wickramasinghe, *Mon. not. R. Astron. Soc.*, **181**, 51P, 1977.

14. "Spectroscopic evidence for interstellar grain clumps in meteoritic inclusions", A. Sakata, N. Nakagawa, T. Iguchi, S. Isobe, M. Morimoto, F. Hoyle and N.C. Wickramasinghe, *Nature*, **266**, 241, 1977.

15. "Polysaccharides and the infrared spectra of galactic sources", F. Hoyle and N.C. Wickramasinghe, *Nature*, **268**, 610, 1977.

16. "Prebiotic polymers and infrared spectra of galactic sources", N.C. Wickramasinghe, F. Hoyle, J. Brooks and G. Shaw, *Nature*, **269**, 674, 1977.

17. "Identification of the 2200 Å interstellar absorption feature", F. Hoyle and N.C. Wickramasinghe, *Nature*, **270**, 323, 1977.

18. "Origin and nature of carbonaceous material in the galaxy", F. Hoyle and N.C. Wickramasinghe, *Nature*, **270**, 701, 1977.

19. "Identification of interstellar polysaccharides and related hydrocarbons", F. Hoyle, N.C. Wickramasinghe and A.H. Olavesen, *Nature*, **271**, 229, 1978.

20. "Calculations of infrared fluxes from galactic sources for a polysaccharide grain model", F. Hoyle and N.C. Wickramasinghe, *Astrophys. Space Sci.*, **53**, 489, 1978.

21. "Comets, ice ages and ecological catastrophes", F. Hoyle and N.C. Wickramasinghe, *Astrophys. Space Sci.*, **53**, 523, 1978.

22. "Biochemical chromophores and the interstellar extinction at ultraviolet wavelengths", F. Hoyle and N.C. Wickramasinghe, *Astrophys. Space Sci.*, **65**, 241, 1979.

23. "On the nature of interstellar grains", F. Hoyle and N.C. Wickramasinghe, *Astrophys. Space Sci.*, **66**, 77, 1979.

24. "The identification of the 3 micron spectral feature in galactic infrared sources", F. Hoyle and N.C. Wickramasinghe, *Astrophys. Space Sci.*, **68**, 499, 1980.

25. "Organic grains in space", F. Hoyle and N.C. Wickramasinghe, *Astrophys. Space Sci.*, **69**, 511, 1980.

26. "Organic material and the 1.5–4 micron spectra of galactic sources", F. Hoyle and N.C. Wickramasinghe, *Astrophys. Space Sci.*, **72**, 183, 1980.

27. "Dry polysaccharides and the infrared spectrum of OH26.5 + 0.6", F. Hoyle and N.C. Wickramasinghe, *Astrophys. Space Sci.*, **72**, 247, 1980.

28. "Evidence for interstellar biochemicals", F. Hoyle and N.C. Wickramasinghe, in *Giant Molecular Clouds in the Galaxy*, (eds.) P.M. Solomon and M.G. Edmunds (Pergamon, 1980).

29. "Why Neo-Darwinism does not work", F. Hoyle and C. Wickramasinghe (University College, Cardiff Press, 1982).

30. "Comets — a vehicle for panspermia", F. Hoyle and N.C. Wickramasinghe (ed.) C. Ponnamperuma (D. Reidel Publishing Co., 1981).

31. "Infrared spectroscopy of micro-organisms near 3.4 microns in relation to geology and astronomy", F. Hoyle, N.C. Wickramasinghe, S. Al-Mufti and A.H. Olavesen, *Astrophys. Space Sci.*, **81**, 489, 1982.

32. "Infrared spectroscopy over the 2.9–3.9 micron waveband in biochemistry and astronomy", F. Hoyle, N.C. Wickramasinghe, S. Al-Mufti, A.H. Olavesen and D.T. Wickramasinghe, *Astrophys. Space Sci.*, **83**, 405–409, 1982.

33. "Interstellar absorptions at $\lambda = 3.3$ and 3.3 microns", S. Al-Mufti, A.H. Olavesen, F. Hoyle and N.C. Wickramasinghe, *Astrophys. Space Sci.*, **84**, 259, 1982.

34. "Organo-siliceous biomolecules and the infrared spectrum of the Trapezium nebula", F. Hoyle, N.C. Wickramasinghe and S. Al-Mufti, *Astrophys. Space Sci.*, **86**, 63, 1982.

35. "A model for interstellar extinction", F. Hoyle and N.C. Wickramasinghe, *Astrophys. Space Sci.*, **86**, 321, 1982.

36. "The infrared spectrum of interstellar dust", F. Hoyle, N.C. Wickramasinghe and S. Al-Mufti, *Astrophys. Space Sci.*, **86**, 341, 1982.

37. "On the optical properties of bacterial grains, I", N.L. Jabir, F. Hoyle and N.C. Wickramasinghe, *Astrophys. Space Sci.*, **91**, 327, 1983.

38. "Interstellar proteins and the discovery of a new absorption feature at $\lambda = 2800$ Å", L.M. Karim, F. Hoyle and N.C. Wickramasinghe, *Astrophys. Space Sci.*, **94**, 223, 1983.

39. "The ultraviolet absorbance spectrum of coliform bacteria and its relationship to astronomy", F. Hoyle, N.C. Wickramasinghe, E.R. Jansz and P.M. Jayatissa, *Astrophys. Space Sci.*, **95**, 227, 1983.

40. "Organic grains in the Taurus interstellar clouds", F. Hoyle and N.C. Wickramasinghe, *Nature*, **305**, 161, 1983.

41. "Bacterial life in space", F. Hoyle and N.C. Wickramasinghe, *Nature*, **306**, 1983.

42. "The spectroscopic identification of interstellar grains", F. Hoyle, N.C. Wickramasinghe and S. Al-Mufti, *Astrophys. Space Sci.*, **98**, 343, 1984.

43. "Proofs that life is cosmic", F. Hoyle and N.C. Wickramasinghe, *Mem. Inst. Fund. Studies*, Sri Lanka, No. 1, 1983.

44. "2.8–3.6 micron spectra of micro-organisms with varying H_2O ice content", F. Hoyle, N.C. Wickramasinghe and N.L. Jabir, *Astrophys. Space Sci.*, **92**, 439, 1983.

45. "The extinction of starlight at wavelengths near 2200 Å", F. Hoyle, N.C. Wickramasinghe and N.L. Jabir, *Astrophys. Space Sci.*, **92**, 433, 1983.

46. "The radiation of microwaves and infrared by slender graphite needles", F. Hoyle, J.V. Narlikar and N.C. Wickramasinghe, *Astrophys. Space Sci.*, **103**, 371, 1984.

47. "The ultraviolet absorbance of presumably interstellar bacteria and related matters", F. Hoyle, N.C. Wickramasinghe and S. Al-Mufti, *Astrophys. Space Sci.*, **111**, 65, 1985.

48. "An object within a particle of extraterrestrial origin compared with an object of presumed terrestrial origin", F. Hoyle, N.C. Wickramasinghe and H.D. Pflug, *Astrophys. Space Sci.*, **113**, 209, 1985.

49. "On the nature of dust grains in the comae of Comets Cernis and Bowell", F. Hoyle, N.C. Wickramasinghe and M.K. Wallis, *Earth, Moon and Planets*, **33**, 179, 1985.

50. "Legionnaires' disease: Seeking a wider cause", F. Hoyle, N.C. Wickramasinghe and J. Watkins, *The Lancet*, 25 May 1985, p. 1216.

51. "Archaeopteryx — a photographic study", R.S. Watkins, F. Hoyle, N.C. Wickramasinghe, J. Watkins, R. Rabilizirov and L.M. Spetner, *British J. Photography* (8 March) **132**, 264, 1985.

52. "Archaeopteryx — a further comment", R.S. Watkins, F. Hoyle, N.C. Wickramasinghe, J. Watkins, R. Rabilizirov and L.M. Spetner, *British J. Photography* (March 29) **132**, 358, 1985.

53. "Archaeopteryx — further evidence", R.S. Watkins, F. Hoyle, N.C. Wickramasinghe, J. Watkins, R. Rabilizirov and L.M. Spetner, *British J. Photography* (April 26) **132**, 468, 1985.

54. "Archaeopteryx — problems arise, and a motive", F. Hoyle and N.C.Wickramasinghe, *British J. Photography* (June 21) **132**, 693, 1985.

55. "The availability of phosphorous in the bacterial model of the interstellar grains", F. Hoyle and N.C. Wickramasinghe, *Astrophys. Space Sci.*, **103**, 189, 1984.

56. "The properties of large particles in the zodiacal cloud and in the interstellar medium and their relation to recent IRAS observations", F. Hoyle and N.C. Wickramasinghe, *Astrophys. Space Sci.*, **107**, 223, 1984.

57. "From grains to bacteria", F. Hoyle and N.C. Wickramasinghe (University College, Cardiff Press, 1984).

58. "Living Comets", F. Hoyle and N.C. Wickramasinghe (University College, Cardiff Press, 1985).

59. "*Viruses from Space*", F. Hoyle and N.C. Wickramasinghe (University College, Cardiff Press, 1986).

60. "On the nature of the interstellar grains", *Q. Jl. R. A. S.*, **27**, 21, 1986.

61. "On the nature of the particles causing the 2200 Å peak in the extinction of starlight", F. Hoyle and N.C. Wickramasinghe, *Astrophys. Space Sci.*, **122**, 181, 1986.

62. "The measurement of the absorption properties of dry microorganisms and its relationship to astronomy", F. Hoyle, N.C. Wickramasinghe and S. Al-Mufti, *Astrophys. Space Sci.*, **113**, 413, 1985.

63. "The viability with respect to temperature of micro-organisms incident on the Earth's atmosphere", F. Hoyle, N.C. Wickramasinghe and S. Al-Mufti, *Earth, Moon and Planets*, **35**, 79, 1986.

64. "Diatoms on Earth, Comets, Europa and in interstellar space", R.B. Hoover, F. Hoyle, N.C. Wickramasinghe, M.J. Hoover and S. Al-Mufti, *Earth, Moon and Planets*, **35**, 19, 1986.

65. "The effects of irregularities of internal structure in determining the ultraviolet extinction properties of interstellar grains", F. Hoyle, N.C. Wickramasinghe, S. Al-Mufti and L.M. Karim *Astrophys. Space Sci.*, **114**, 303, 1985.

66. "The case for interstellar micro-organisms", F. Hoyle, N.C. Wickramasinghe and S. Al-Mufti, *Astrophys. Space Sci.*, **110**, 401, 1985.

67. "Some evidence against the authenticity of Archaeopteryx Lithographica", F. Hoyle, N.C. Wickramasinghe, L.M. Spetner and M. Magaritz, *Bild der Wissenschaft* **5**, 51, 1988.

68. "Interstellar extinction by organic grain clumps", F. Hoyle and N.C. Wickramasinghe, *Astrophys. Space Sci.*, **140**, 191, 1988.

69. "Polymeric complexes in comets and in space", F. Hoyle and N.C. Wickramasinghe, *Astrophys. Space Sci.*, **141**, 177, 1988.

70. "Cosmic Life Force", F. Hoyle and N.C. Wickramasinghe (J.M. Dent, 1988).

71. "A diatom model of dust in the Trapezium nebula", Q. Majeed, N.C. Wickramasinghe, F. Hoyle and S. Al-Mufti, *Astrophys. Space Sci.*, **140**, 205, 1988.

72. "Mineral Grains in the 10 and 20 μm spectral features in the Trapezium nebula", F. Hoyle, N.C. Wickramasinghe and Q. Majeed, *Astrophys. Space Sci.*, **141**, 399, 1988.

73. "Archaeopteryx — more evidence of a forgery", F. Hoyle, N.C. Wickramasinghe, L.M. Spetner and M. Magaritz, *British J. Photography*, pp. 14–18 (7 Jan. 1988).

74. "The infrared excess from the White Dwarf star G29–38: a Brown Dwarf or dust?", F. Hoyle, N.C. Wickramasinghe and S. Al-Mufti, *Astrophys. Space Sci.*, **143**, 193, 1988.

75. "Metallic particles in astronomy", F. Hoyle and N.C. Wickramasinghe, *Astrophys. Space Sci.*, **147**, 245–256, 1988.

76. "The organic nature of cometary grains", N.C. Wickramasinghe, F. Hoyle, M.K. Wallis and S. Al-Mufti, *Earth, Moon and Planets*, **40**, 101, 1988.

77. "Mineral and organic particles in Astronomy", N.C. Wickramasinghe, F. Hoyle and Q. Majeed, *Astrophys. Space Sci.*, **158**, 335, 1989.

78. "Modelling the 5–30 μm spectrum of Comet Halley", N.C. Wickramasinghe, M.K. Wallis and F. Hoyle, *Earth, Moon and Planets*, **43**, 145, 1988.

79. "Aromatic hydrocarbons in very small interstellar grains", N.C. Wickramasinghe, F. Hoyle and T. Al-Jubory, *Astrophys. Space Sci.*, **158**, 135, 1989.

80. "An integrated 2.5–12.5 μm emission spectrum of naturally occurring aromatic molecules", N.C. Wickramasinghe, F. Hoyle and T. Al-Jubory, *Astrophys. Space Sci.*, **166**, 333, 1990.

81. "Extraterrestrial particles and the greenhouse effect", N.C. Wickramasinghe, F. Hoyle and R. Rabilizirov, *Earth, Moon and Planets*, **46**, 297, 1989.

82. "Greenhouse dust", N.C. Wickramasinghe, F. Hoyle and R. Rabilizirov, *Nature*, **341**, 28, 1989.

83. "A unified model for the 3.28 μm and the 2200 Å interstellar extinction feature", F. Hoyle and N.C. Wickramasinghe, *Astrophys. Space Sci.*, **154**, 143, 1989.

84. "Linear and circular polarization by hollow organic grains", F. Hoyle and N.C. Wickramasinghe, *Astrophys. Space Sci.*, **151**, 285, 1989.

85. "The microwave background in steady-state cosmology", F. Hoyle and N.C. Wickramasinghe, *ESA SP-290*, 489, 1989.

86. "A unified model for the 3.28 μm and 3.4 μm spectral feature in the interstellar medium and in comets", F. Hoyle and N.C. Wickramasinghe, *ESA SP-290*, 67, 1989.

87. "Biologic versus abiotic models of cometary dust", M.K. Wallis, N.C. Wickramasinghe, F. Hoyle and R. Rabilizirov, *Mon. not. R. Astron. Soc.*, **238**, 1165–1170, 1989.

88. "The extragalactic Universe: and alternative view", H.C. Arp, G. Burbidge, F. Hoyle, J.V. Narlikar and N.C. Wickramasinghe, *Nature*, **346**, 807–812, 1990.

89. "The case for life as a cosmic phenomenon", F. Hoyle and N.C. Wickramasinghe, *Nature*, **322**, 509, 1986.

90. "Sunspots and influenza", F. Hoyle and N.C. Wickramasinghe, *Nature*, **343**, 304, 1990.

91. "Influenza — evidence against contagion: discussion paper", F. Hoyle and N.C. Wickramasinghe, *J. Roy. Soc. Med.*, **83**, 258, 1990.

92. "The microwave background: its smoothness and frequency distribution as an astrophysical product", F. Hoyle, N.C. Wickramasinghe and G. Burbidge, *29th Liege International Astrophysical Colloquium*, July 2–6, 1990.

93. "Mineral grains in interstellar space", N.C. Wickramasinghe, F. Hoyle, S. Al-Mufti and T. Al-Jabory, in *Dusty Objects in the Universe*, (eds.) E. Bussoletti and A.A. Vittone (Kluwer Academic Press, 1990).

94. "Back-scattering of sunlight by ice grains in the Mesosphere", F. Hoyle and N.C. Wickramasinghe, *Earth, Moon and Planets*, **52**, 161–170, 1991.

95. "The implications of life as a cosmic phenomenon: The anthropic context", F. Hoyle and N.C. Wickramasinghe, *J. British Interplan. Soc.*, **44**, 77–86, 1991.

96. "Cometary habitats for primitive life", M.K. Wallis, N.C. Wickramasinghe and F. Hoyle, *Adv. Space Res.*, **12**(4), 281–285, 1992.

97. "The extinction of starlight revisited", N.C. Wickramasinghe, B. Jazbi and F. Hoyle, *Astrophys. Space Sci.*, **186**, 67–80, 1991.

98. "Extinction properties of infinitely long graphite cylinders", B. Jazbi, F. Hoyle and N.C. Wickramasinghe, *Astrophys. Space Sci.*, **186**, 151–155, 1991.

99. "The case against graphite particles in interstellar space", N.C. Wickramasinghe, A.N. Wickramasinghe and F. Hoyle, *Astrophys. Space Sci.*, **196**, 167–169, 1992.

100. "The absorption of electromagnetic radiation by metal cylinders of finite length", N.C. Wickramasinghe, A.N. Wickramasinghe and F. Hoyle, *Astrophys. Space Sci.*, **193**, 141–144, 1992.

101. "Comets as a source of interplanetary and interstellar grains", F. Hoyle and N.C. Wickramasinghe, in *Origin and Evolution of Interplanetary Dust* (eds.) A.C. Levasseur-Regourd and H. Hasegawa (Kluwer Academic Publishers, 1991), pp. 235–240.

102. "Microdiamonds and the 3.4 micron feature in protostellar sources", F. Hoyle and N.C. Wickramasinghe, *Astrophys. Space Sci.*, **207**, 309–311, 1993.

103. "Absorption properties of astronomical iron whiskers: an accurate crogenic model", N.C. Wickramasinghe and F. Hoyle, *Astrophys. Space Sci.*, **213**, 143–154, 1994.

104. "Critique of Fischer-Tropsch type reactions in the solar nebula", S. Ramadurai, F. Hoyle and N.C. Wickramasinghe, *Bull. Astron Soc. India* **21**, 329–334, 1993.

105. "Biofluorescence and the extended red emission in astrophysical sources", F. Hoyle and N.C. Wickramasinghe, *Astrophys. Space Sci.*, **235**, 343–347, 1996.

106. "Very small dust grains (VSDP's) in Comet C/1996 B2 (Hyakutake)", N.C. Wickramasinghe and F. Hoyle, *Astrophys. Space Sci.*, **239**, 121, 1996.

107. "Eruptions from comet Hale-Bopp at 6.5AU", N.C. Wickramasinghe, F. Hoyle and D. Lloyd, *Astrophys. Space Sci.*, **240**, 161, 1996.

108. "Infrared signatures of prebiology — or biology", N.C. Wickramasinghe, F. Hoyle, S. Al-Mufti and D.H. Wallis, in *Astronomical and Biochemical Origins and the Search for Life in the Universe* (eds.) C.B. Cosmovici, S. Bowyer and D. Werthimer (Editrice Compositori, 1997).

109. "Comet P/Shoemaker-Levy 9 collision with Jupiter: a model of G-site dust composition", D.H. Wallis and N.C. Wickramasinghe, *Astrophys. Space Sci.*, **254**, 25–35, 1997.

110. "Spectroscopic evidence for panspermia", N.C. Wickramasinghe, F. Hoyle and D.H. Wallis, *Proc. SPIE*, **3111**, 282–295, 1997.

111. "The astonishing redness of Kuiper-Belt objects", N.C. Wickramasinghe and F. Hoyle, *Astrophys. Space Sci.*, **259**, 205–208, 1998.

112. "Microdiamonds and the ultraviolet extinction of starlight", *Astrophys. Space Sci.*, **259**, 379–383, 1998.

113. "Infrared evidence for panspermia: an update", *Astrophys. Space Sci.*, **259**, 385–401, 1998.
    Miller-Urey synthesis in the nuclei of galaxies", N.C. Wickramasinghe and F. Hoyle, *Astrophys. Space Sci.*, **259**, 99–103, 1998.

114. "Search for living cells in stratospheric samples", J.V. Narlikar, S. Ramadurai, P. Bhargava, S.V. Damle, N.C. Wickramasinghe, D. Lloyd, F. Hoyle and D.H. Wallis, *Proc. SPIE*, **3441**, 301–305, 1998.

115. "Panspermia in perspective", N.C. Wickramasinghe, F. Hoyle and B. Klyce, *Proc. SPIE*, **3441**, 306–318, 1988.

116. "Cosmological panspermia", N.C. Wickramasinghe and F. Hoyle. *Proc. SPIE*, **3441**, 319–323, 1998.

117. "Towards an understanding of the nature of racial prejudice", F. Hoyle and N.C. Wickramasinghe, *J. Scientific Exploration*, **13**, 681–684, 1999.

118. "Cosmic Life: Evolution and Chance", F. Hoyle and N.C. Wickramasinghe, *The Biochemist*, **21**(6), 1999.

119. "Astronomical Origins of Life: Steps towards Panspermia", F. Hoyle and N.C. Wickramasinghe (Kluwer Academic Publishers, 2000).
120. "Cross-linked Heteroaromatic Polymers in Interstellar dust", N.C. Wickramasinghe, D.T. Wickramasinghe and F. Hoyle, *Astrophys. Space Sci.*, **275**, 181–184, 2001.
121. "A bacterial "singerprint in a Leonid meteor train", N.C. Wickramasinghe and F. Hoyle, *Astrophys. Space Sci.*, **277**, 625–628, 2001.
122. "The detection of living cells in stratospheric samples", Melanie J. Harris, N.C. Wickramasinghe, David Lloyd, M. Turner, F. Hoyle, J.V. Narlikar and P. Rajaratnam, *Proc. SPIE*, **4495**, 192–198, 2002.

Books

Lifecloud: The Origin of Life in the Galaxy: F. Hoyle and N.C. Wickramasinghe (J.M. Dent, Lond., 1978).

Diseases from Space: F. Hoyle and N.C. Wickramasinghe (J.M. Dent, Lond., 1979).

Origin of Life: F. Hoyle and N.C. Wickramasinghe (University College Cardiff Press, 1979).

Space Travellers: The Bringers of Life: F. Hoyle and N.C. Wickramasinghe (University College Cardiff Press, 1981).

Evolution from Space: F. Hoyle and N.C. Wickramasinghe (J.M. Dent, 1981).

Is Life an Astronomical Phenomenon?: F. Hoyle and N.C. Wickramasinghe (Universtiy College, Cardiff Press, 1982).

Why Neo Darwinism does not Work: F. Hoyle and N.C. Wickramasinghe (University College Cardiff Press, 1982).

Proofs that Life is Cosmic: F. Hoyle and N.C. Wickramasinghe (Inst. of Fund. Studies, Sri Lanka, Mem, No. 1, 1982).

From Grains to Bacteria: F. Hoyle and N.C. Wickramasinghe (University College, Cardiff Press, 1984).

Living Comets: F. Hoyle and N.C. Wickramasinghe (University College, Cardiff Press, 1985).

Viruses from Space: F. Hoyle and N.C. Wickramasinghe (University College Cardiff Press, 1986).

Archaeopteryx — The Primordial Bird: A Case of Fossil Forgery: F. Hoyle and N.C. Wickramasinghe (Christopher Davies, Swansea, 1986).

Cosmic Life Force: F. Hoyle and N.C. Wickramasinghe (J.M. Dent, Lond., 1988).

The Theory of Cosmic Grains: F. Hoyle and N.C. Wickramasinghe (Kluwer Academic Publishers, 1990).

Our Place in the Cosmos: F. Hoyle and N.C. Wickramasinghe (Weidenfeld and Nicholson, Lond., 1993).

Life of Mars: The Case for a Cosmic Heritage: F. Hoyle and N.C. Wickramasinghe (Clinical Press, 1997).

Astronomical Origins of Life: Steps Towards Panspermia: F. Hoyle and N.C. Wickramasinghe (Kluwer Academic Press, 2000).

Bibliography to Second Edition

Technical Papers

1. "Horizontal gene transfer in evolution: facts and challenges", L. Boto, *Proc. R. Soc. B* 2010, **277**, 819–827, 2009.
2. "Evidence for an ancient selective sweep in the MHC class I gene repertoire of chimpanzees", N.G. De Groot, N. Otting, G.G.M. Doxiadis *et al.*, *PNAS*, **99**, 11748–11753, 2002.
3. "Dust extinction in high-z galaxies with gamma-ray burst afterglow spectroscopy: The 2175 Å feature at $z = 2.45$", Á. Elíasdóttir, J.P.U. Fynbo, J. Hjorth *et al.*, *Astrophysical Journal*, **697**, 1725–1740, 2009.
4. "Turbulence in the ocean, atmosphere, galaxy and universe", C.H. Gibson, *Appl. Mech. Rev.*, **49**, 299–315, 1996.
5. "Turbulent formation of protogalaxies at the end of the plasma epoch: Theory and observations", C.H. Gibson and R.E. Schild, *J. Cosmol.*, **6**, 1351–1360, 2010.
6. "The origin of life from primordial planets", C.H. Gibson, R.E. Schild and N.C. Wickramasinghe, *Int. J. Astrobiol.*, **10**(2), 83–98, 2011.
7. "Creation of a bacterial cell controlled by a chemically synthesised genome", D.G. Gibson, J.I. Glass, C. Lartigue *et al.*, *Science*, **329**, 52–56, 2010.
8. R.B. Hoover, in *Perspectives in Astrobiology*, (eds.) R.B. Hoover, A.Y. Rozanov and R.R. Paepe (Amsterdam: IOS press, 2005) p. 366, 43.
9. "Fossils of cyanobacteria in CII carbonaceous meteorites", R.B. Hoover, *Journal of Cosmology*, **13**, 2011.
10. "Endogenous non-retroviral RNA virus elements in mammalian genomes", M. Horie, T. Honda, Y. Suzuki *et al.*, *Nature*, **463**, 84–87, 2010.

11. "Ocean-lie Water in the Jupiter-Family Comet Hartley 2", P. Hartogh, D.C. Lis, D. Bockelee-Morva *et al.*, *Nature*, **476**, 218–220, 2011.
12. "On graphite particles as interstellar grains", F. Hoyle and N.C. Wickramasinghe, *Mon. Not. R. Astr. Soc.*, **124**, 417–433, 1962.
13. "Influenza — evidence against contagion", F. Hoyle and N.C. Wickramasinghe, *J. Roy. Soc. Med.*, **83**, 258–261, 1990.
14. R. Jain, M.C. Rivera, J.E. Moore *et al.*, *Mol. Biol. Evol.*, **20**(10), 1598–1602, 2003.
15. R. Joseph and N.C. Wickramasinghe, *Journal of Cosmology*, 2011.
16. "Horizontal gene transfer in eukaryotic evolution", P.J. Keeling and J.D. Palmer, *Nature Reviews Genetics*, **9**, 605–618, 2008.
17. "Chemical properties in the most distant radio galaxy", K. Matsuoka, T. Nagao, R. Mailino *et al.*, *Astron. Astrophys.*, **532**, L10, 2011.
18. "Detection of the 2175 Å extinction feature at $z = 0.83$", V. Motta, E. Mediavilla, J.A. Muñoz *et al.*, *ApJ.*, **574**, 719–725, 2002.
19. "A mechanism for interstellar panspermia", W.M. Napier, *Mon. Not. R. Astr. Soc.*, **348**, 46–51, 2004.
20. "Diffuse molecular gas at high redshift — detection of CO molecules and the 2175 Å dust feature at $z = 1.64$", P. Noterdaeme, C. Ledoux, R. Srianand *et al.*, *Astronomy & Astrophysics*, 2009.
21. "Evidence for biodegradation products in the interstellar medium", K. Rauf and C. Wickramasinghe, *Int. J. Astrobiol.*, **9**(1), 29–34, 2010.
22. "*Janibacter hoylei* sp.nov., *Bacillus isronensis* sp.nov. and *Bacillus aryabhattai* sp.nov. isolated from cryotubes used for collecting air from the upper atmosphere", S. Shivaji, P. Chaturvedi, Z. Begum *et al.*, *Int. J. Systematic and Evolutionary Microbiology*, **59**, 2977–2986, 2009.
23. "Measuring PAH emission in ultradeep Spitzer IRS spectroscopy of high-red-shift IR luminous galaxies", H.I. Teplitz, V. Desai, L. Armuo *et al.*, *ApJ.*, **659**, 941–949, 2007.
24. "The sequence of the human genome", J.C.J. Venter, M.D. Adams, E.W. Myers *et al.*, *Science*, **291**, 1304–1351, 2001.
25. "Microorganisms cultured from stratospheric air samples obtained at 41 km", M. Wainwright, N.C. Wickramasinghe, J.V. Narlikar and P. Rajaratnam, *FEMS Microbiology Letters*, **218**, 161, 2003.
26. "Discovery of biological structures in the Tissint Mars meteorite", J. Wallis, C. Wickramasinghe, D. Wallis *et al.*, *J. Cosmol.*, **18** (http://journalofcosmology.com/JOC18/TissintFinal.pdf), 2012.
27. "Interstellar transfer of planetary microbiota", M.K. Wallis and N.C. Wickramasinghe, *Mon. Not. R. Astr. Soc.*, **348**, 52–57, 2004.
28. "Specific inactivation of two immunomodulatory SIGLEC genes during human evolution", X. Wang, N. Mitra, I. Secundino *et al.*, *PNAS* Early Edition, doi/10.1073/pnas.1119459109, 2012.

29. "Panspermia, past and present: Astrophysical and biophysical conditions for the dissemination of life in space", P. Wesson, *Sp. Sci. Rev.*, **156**(1–4), 239–252, 2010.

30. N.C. Wickramasinghe, F. Hoyle and D. Lloyd, *Astrophys. Sp. Sci.*, **240**, 161, 1996.

31. "Life-bearing planets in the solar vicinity", N.C. Wickramasinghe, J. Wallis, D.H. Wallis, R.E. Schild and C.H. Gibson, *Astrophys. Sp. Sci.*, **341**, 295–299, 2011.

32. "The astrobiological case for our cosmic ancestry", N.C. Wickramasinghe, *Int. J. Astrobiol*, **9**(2), 119–129, 2010.

33. "DNA sequencing and predictions of the cosmic theory of life", N.C. Wickramasinghe, *Astrophys. Sp. Sci.*, DOI 10.1007s10509-012-1227-y, 2012.

Books

Viruses from Space: F. Hoyle, C. Wickramasinghe and J. Watkins (Univ. Coll. Cardiff Press, 1986).

Diseases from Space: F. Hoyle and N.C. Wickramasinghe (J.M. Dent & Sons, Lond., 1979).

Evolution from Space: F. Hoyle and N.C. Wickramasinghe (J.M. Dent & Sons, Lond., 1981).

Proofs that Life is Cosmic: F. Hoyle and N.C. Wickramasinghe (Mem. Inst. Fund. Studies, Sri Lanka, Vol. 1, No. 1, 1981).

Comets and the Origin of Life: J.T. Wickramasinghe, N.C. Wickramasinghe and W.M. Napier (World Scientific, Singapore, 2010).

Index